Diabetes and Lipids
Second edition

GW00691783

Dr John Reckless is Consultant Endocrinologist at the Royal United Hospital, Bath, and Honorary Reader in Medicine, and in Biochemistry, at the University of Bath. Dr Reckless has clinical and research interests in diabetes and its clinical complications. This is especially in large artery disease and its causes and management, and in abnormalities of lipoprotein metabolism in diabetes.

Diabetes and Lipids
Second edition

JPD Reckless

Consultant Endocrinologist
Department of Endocrinology and Diabetes
Royal United Hospital, Bath

Honorary Reader in
Postgraduate Medicine and in Biochemistry
University of Bath
Bath, UK

MARTIN DUNITZ

© Martin Dunitz Ltd 1994, 2001

The right of John PD Reckless to be
identified as author of this work has been
asserted by him in accordance with the
Copyright, Designs and Patents Act, 1988

First published in the United Kingdom
in 1994 by
Martin Dunitz Ltd
7– 9 Pratt Street
London NW1 0AE

Tel: +44 (0)207 482 2202
Fax: +4 (0)207 267 0159
Email: info@dunitz.co.uk
Website: http://www.dunitz.co.uk

First edition 1994
Second edition 2001

Although every effort has been made to ensure
that drug doses and other information are
presented accurately in this publication, the ultimate
responsibility rests with the prescribing physician.
Neither the publishers nor the authors can be held
responsible for errors for any consequences arising
from the use of information contained herein. For
detailed prescribing information or instructions on
the use of any product or procedure discussed
herein, please consult the prescribing information
or instructional material used by the manufacturer.

A CIP Record for this book is available
from the British Library.

ISBN 1-85317-614-1

Distributed in the USA, Canada and Brazil by:

Blackwell Science Inc.
Commerce Place, 350 Main Street
Malden MA 02148, USA
Tel: 1 800 215 1000

Printed and bound in Italy by Printer Trento S.r.l.

Contents

Glossary

ACAT	acyl cholesterol acyltransferase
ACE	angiotensin-converting enzyme
apo	apoprotein
CETP	cholesterol ester transfer proteins
CHD	coronary heart disease
CVD	cerebrovascular disease
FCH	familial combined hyperlipidaemia
HDL	high-density lipoproteins
HbA_{1c}	glycosylated haemoglobin
HL	hepatic lipase
HMG-CoA	3-hydroxy-3-methylglutaryl coenzyme A
IDDM	insulin-dependent diabetes mellitus (Type 1 diabetes)
IDL	intermediate-density lipoproteins
IGT	impaired glucose tolerance
LCAT	lecithin:cholesterol acyltransferase
LDL	low-density lipoproteins
Lp(a)	lipoprotein (a)
LPL	lipoprotein lipase
LRP	LDL receptor-related protein
mRNA	messenger RNA
NIDDM	non-insulin-dependent diabetes mellitus (Type 2 diabetes)
PAI-1	plasminogen activator inhibitor-1
PVD	peripheral vascular disease
UKPDS	United Kingdom Prospective Diabetes Study
VLDL	very-low-density lipoproteins

Introduction

In diabetes mellitus, the major cause of mortality and a common cause of morbidity is macrovascular or large artery disease, in the forms of coronary heart disease (CHD), stroke and cerebrovascular disease (CVD) and peripheral vascular disease (PVD). Diabetes is a generalized disorder of metabolism and not just of glucose. This book considers lipids and lipoproteins in diabetes, and their relationship to short-term and long-term symptoms and complications of diabetes such as macrovascular disease. Relationships with other macrovascular disease risk factors are important for they influence management strategies. Therapeutic options for both lifestyle change and drug interventions are considered, providing an up-to-date view of current knowledge and practice.

- Diabetes mellitus is a disorder of carbohydrate, fat and protein metabolism, characterized by relative or absolute insulin lack.

Coronary heart disease and diabetes

Acute and chronic complications occur in insulin-dependent (Type 1 diabetes or IDDM) and non-insulin-dependent (Type 2 diabetes or NIDDM) diabetes mellitus, and may be accompanied by lipid disorders and changes in lipid metabolism. Although most acute symptoms are limited to glucose metabolism, lipid handling is often deranged, with high levels of fatty acids, ketones and hypertriglyceridaemia. Severe hypertriglyceridaemia can cause acute pancreatitis.

Chronic complications of diabetes may be specific to diabetes (small vessel or microvascular disease) or reflect an increased risk of problems seen in the non-diabetic population (large vessel or macrovascular disease). The two can be related, because, for example, CHD is greatly increased in diabetic renal disease with proteinuria. Both are related to duration of diabetes; in addition, microvascular disease is clearly related to quality of diabetic control.[1–4]

There is a blood sugar threshold for microvascular complications (retinopathy, neuropathy, nephropathy) which occur in diabetes, but not in impaired glucose tolerance (IGT). However, macrovascular disease is increased in both diabetes and IGT; it is also doubled in normal individuals whose blood glucose levels are in the top 5% of population values.[5] High levels of blood glucose and glycosylated haemoglobin (HbA_{1c}), in non-diabetic individuals are associated with increased cardiovascular disease.[6,7]

Other factors, in addition to glycaemic control, may also influence macrovascular disease. Hypertension, dyslipidaemia (high levels of triglycerides, low levels of high-density lipoprotein (HDL) cholesterol) and hyperfibrinogenaemia are more likely to be present over many years, before and after diabetes diagnosis.

Excess risk of atherosclerosis and CHD in Type 1 diabetes and Type 2 diabetes

Atherosclerotic lesions are similar in both diabetic and non-diabetic patients, but they are more diffuse and extensive when associated with diabetes; in such patients the iliofemoral, popliteal and tibial vessels are affected more. Diffuse diabetic disease results in the finding of multiple lesions and narrowing in the coronary arteries at clinical presentation, so surgery or angioplasty may be more difficult, with a higher mortality after coronary thrombosis. Autonomic neuropathy could explain the presence of more silent ischaemia and atypical angina which lead to a later clinical presentation.

Coronary heart disease

After 1922, when death rates in younger diabetic patients were reduced by giving insulin, macrovascular disease became the major cause of mortality (Figure 1). Three-quarters of diabetic patients in Western populations now die from macrovascular disease, and half from CHD (Figure 2).

Among the Japanese, in whom CHD is uncommon, CHD is also uncommon in diabetic patients, but still increased compared to

Figure 1
Changing patterns of mortality in diabetes over 70 years. Since the introduction of insulin, rates of death from acute complications have reduced considerably, but chronic, largely macrovascular, causes have become predominant. Data from Marble.[8]

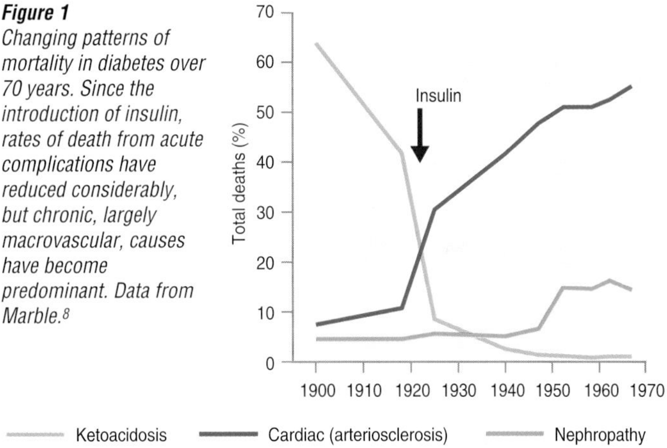

the non-diabetic population.[9] CHD in diabetic Japanese people is on the increase, and diabetic Japanese migrants to Hawaii have CHD rates that approach those of diabetic Caucasians.[10]

In a study from the Joslin Clinic, Boston, mortality rates were 1.5–7-fold higher in men, and 2.5–14-fold higher in women, compared with the general population; the maximum relative risk was at age 30 years, but of course the absolute risk is much greater in the older Type 2 population. Life expectancy was reduced by 30% at all ages of diabetes onset (Figure 3). The general fall in CHD death rates after 1968 in the US has not been as apparent in people with diabetes. A three- to five-fold increased CHD risk (greater in females than males) in diabetes compared to non-diabetic populations has been demonstrated in many subsequent studies. Besides a higher CHD risk, individuals with diabetes are more likely to die from their first event within the first 24 hours, and at 1 month or 1 year of follow-up. Compared to non-diabetic people, they have a worse prognosis, higher recurrence rate and more difficult

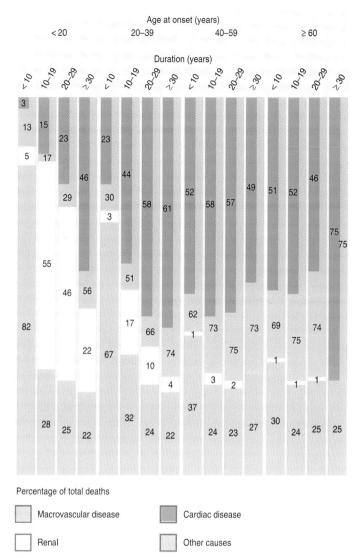

Percentage of total deaths

Macrovascular disease   Cardiac disease

Renal   Other causes

Figure 2
Relationships of causes of death in diabetic patients with the duration of diabetes and age at diagnosis. Macrovascular disease increases with both age and diabetes duration, and cardiac disease accounts for the majority of the total vascular disease cases. Data from Marks and Krall.[13]

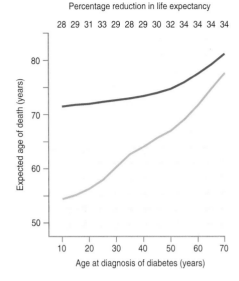

Figure 3
Life expectation by age at onset of diabetes, compared to non-diabetic controls. Percentage reduction in life expectancy is shown. Deaths within one week of first observation or hospital discharge are excluded. Data from the Joslin Clinic analysed by the Metropolitan Life Insurance Company; data from Marks and Krall.[13]

Diabetes
Non-diabetes

revascularization procedures with poorer results. Reductions in CHD rates in non-diabetic populations over recent years have not been mirrored in diabetic groups.

Cerebrovascular disease

As for CHD, age-adjusted rates for cerebrovascular disease were 2.5–3.5 times higher in diabetic patients when compared with controls in the Framingham Study and in the UK.[11,12]

Peripheral vascular disease

Diabetic patients have more PVD than non-diabetic patients; PVD is particularly increased when hypertension and/or smoking coexist.[14] Aortic and medial calcification in leg arteries is more frequent. Claudication is four times more common in men with diabetes, and six times more common in women with diabetes compared to those without.[15,16] Arterial occlusion below the knee is increased, and gangrene is 60 times more common.[12]

Amputation rates rise with age, but the relative risk is at its greatest below the age of 45 years. Diabetic neuropathic and microvascular changes contribute to this, but atheroma is a major feature. Half of all amputations occur in diabetic patients,[17] and relate to the duration of diabetes.[18] The causes of PVD and CHD are similar with smoking and diabetes being predominant; however, strategies for CHD prevention should influence PVD. The distribution of diabetic vascular disease, together with some of the lipid changes, are quite similar to those occurring in remnant lipaemia (familial dysbetalipoproteinaemia; Type III hyperlipidaemia) in non-diabetic patients.

CHD risk factors and diabetes

Three main modifiable CHD risk factors are hypertension, cigarette smoking and dyslipidaemia, with diabetes itself being a fourth factor.[19] The relative protection of women, especially before the menopause, is lost in diabetes.

Poor diabetes control might increase CHD risk, and some improvement in risk is probably seen with improved diabetic control,[3,4] but is unproven.[20] The definition of diabetes is arbitrary, linked to a blood glucose level above which microangiopathy occurs; however, a consuming interest in normoglycaemia and microvascular disease has led to insufficient attention being paid to other metabolic changes (such as lipid and lipoprotein abnormalities) relevant to macrovascular disease, that occur at lower glucose levels. Ideal diabetic control should also normalize triglyceride and HDL cholesterol levels, and ameliorate insulin resistance and hyperinsulinaemia but only partly does so. Clustering of conventional risk factors only partly accounts for excess risk in diabetes.

Individual CHD risk factors and diabetes

Main modifiable risk factors for CHD in diabetes
- Smoking
- Hypertension
- Hyperlipidaemia

Smoking

Smoking is associated with excess CHD, particularly in high-risk groups[19] (Figure 4); cessation of smoking is therefore essential, as smoking could double the 10-year mortality.

Hypertension

The prevalence of hypertension doubles in diabetes; it is present in 50% of Type 1 diabetes and 65% of Type 2 diabetes patients after 30 years' duration of diabetes. Sodium retention may increase vascular tone and contribute to nephropathy. Hypertension is associated with four-fold mortality from CHD and stroke (Figure 4), particularly in women and Afro-Caribbean people. In Type 2 diabetes with central adiposity, hypertension often accompanies hyperinsulinaemia, insulin resistance and hyperlipidaemia.

Hyperlipidaemia

Diabetes is primarily associated with increased triglyceride levels and decreased HDL cholesterol levels; for a given cholesterol [or low density lipoprotein (LDL) cholesterol] level, the cardiovascular risk is higher for the diabetic than for the non-diabetic patient (Figure 4).

Figure 4

Deaths from cardiovascular disease over 12 years, and the effects of cigarette smoking, hypercholesterolaemia and hypertension in diabetic and non-diabetic men from the screenees of the multiple risk factor intervention trial. Data from Stamler et al.[19]

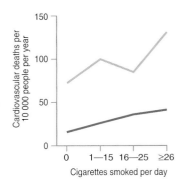

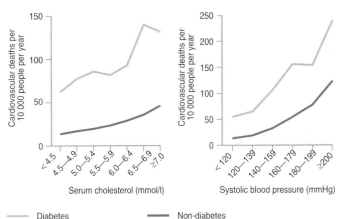

Triglyceride is a more powerful univariate risk factor for CHD but not after multivariate analysis; however, such analysis is not entirely appropriate because triglycerides are not independent of cholesterol, low HDL levels, obesity, diabetes or hypertension. Trials have not been designed with triglycerides as the primary endpoint, although those hypertriglyceridaemic patients receiving gemfibrozil in the Helsinki Heart Study benefited most,[21] including the small diabetic subgroup.[22]

The recent VA-HIT study targeting HDL cholesterol change using gemfibrozil showed a 22% reduction in events overall ($P = 0.006$) while in the diabetic male subgroup the reduction was similar at 25% ($P = 0.05$).[23] In diabetes, hypertriglyceridaemic lipoproteins, in particular, have the potential to be atherogenic, without necessarily being accompanied by severe hypercholesterolaemia.[24]

Diabetic patients with high cholesterol or LDL cholesterol, or low HDL cholesterol levels have more vascular disease.[25] Vascular disease in men was linked to serum or LDL cholesterol (and to triglyceride and very-low-density lipoprotein (VLDL) triglyceride levels) but not to low HDL levels. However, in women, vascular disease had an inverse correlation with HDL cholesterol level, and a positive but weaker correlation with triglyceride, VLDL triglyceride and LDL cholesterol levels. In Type 2 diabetes, vascular disease had a negative correlation with HDL cholesterol levels. In Type 1 diabetes there was no HDL link, but there were positive relationships of CHD with cholesterol, LDL cholesterol, triglyceride and VLDL triglyceride levels. In diabetic women (but not men) in the Framingham Study, HDL level showed an inverse correlation with CHD.

Overweight

Central adiposity exacerbates hypertension, dyslipidaemia and hyperinsulinaemia. Low HDL levels in Type 2 diabetes are associated with obesity and insulin resistance. Truncal or male pattern adiposity over the abdomen, as opposed to gynaecoid distribution over the thighs and hips, is more common in diabetic women than in non-diabetic, and may contribute to their increased CHD risk.

Exercise

Exercise will decrease insulin secretion, increase insulin action and decrease insulin resistance. In thin, insulin-sensitive individuals exercise may not be especially helpful, but the overweight, unfit, insulin-resistant, Type 2 diabetes individual should benefit.

Diabetic renal disease and lipids

Proteinuria in Type 1 diabetes is a major risk factor for early death from renal failure and CHD. Proteinuria increases age-adjusted mortality in diabetes by four- to 36-fold,[26] and is linked to increased dyslipidaemia and accelerated hypertension. Lipoprotein quality is abnormal and all risk factors should be treated energetically. While there are other causes of proteinuria, and while it is a less precise surrogate measure for diabetic

- Hypertension
- Hypercholesterolaemia
- Hypertriglyceridaemia
- Low HDL levels
- Insulin resistance
- Smoking
- Obesity
- Hyperfibrinogenaemia
- Increased PAI-1

renal disease in Type 2 diabetes than in Type 1 diabetes, it is still linked to premature death and morbidity.[27] Furthermore, there is some evidence that lowering cholesterol may retard progression of diabetic nephropathy.[28]

Multiple risk factors in diabetes

The following may cluster together with diabetes or IGT:

Multiple risk factors lead to a cumulative CHD risk which directs treatment priorities (see Figures 10 and 11).

Current concepts in atherogenesis

Understanding of atherogenesis comes from epidemiological, clinical and postmortem studies, animal models, experimental pathology and cell biochemistry. Atheroma occurs in muscular (carotid and coronary) and fibroelastic (aortoiliac) arteries, and arterialized vein grafts. Early fatty streaks progress to complicated lesions, in which a fibrous plaque caps a largely acellular core. A thin cap may fracture or ulcerate, exposing thrombogenic constituents of a large, necrotic, lipid-rich core. Plaque lipid largely derives from LDL – and perhaps lipoprotein(a)(Lp(a)) – which are modified and trapped on subintimal proteoglycans. Monocytes traverse the vessel wall in response to chemotactic factors. They are transformed into macrophages, express oxidized-LDL receptors and take up modified LDL to become foam cells. Endothelial damage, activated macrophages and high LDL levels are associated with release of growth factors; these growth factors encourage smooth muscle cell proliferation, and collagen and proteoglycan synthesis.

Arterial damage and repair are continuous; the factors that favour progression are those that increase endothelial injury or lipid infiltration. Plaque initiation requires cell proliferation and lesion progression requires lipid involvement. Endothelial injury may be mechanical, hypoxic, hormonal or metabolic. Lipoproteins may be glycated, oxidized or minimally modified by endothelial cells. Some factors lead to endothelial damage and platelet aggregation, whereas others lead to abnormal ingress, quantity, quality or immunogenicity of lipoproteins.

Many coronary thrombotic events occur in relation to moderate plaques that are asymptomatic, rather than on blood flow-limiting, > 85% occluded, lesions causing angina. This has led to the concept of the vulnerable plaque. Cytokine production by macrophages may encourage breakdown of smooth muscle cell-produced matrix collagen and glycosoaminoglycans, and weaken the shoulder of the atherosclerotic plaque cap. Lipid lowering may reduce macrophage formation and increase plaque stability.[29]

Present therapy addresses avoidance of endothelial damage (eg anti-smoking and hypotensive measures), lipid lowering, and anti-thrombotic measures (eg low-dose aspirin). In the future, endothelial protection, limitation of lipoprotein oxidation, roles of nitric oxide and free radicals, modification of monocyte–macrophage activation and plaque stabilization will become more important.

Lipids and lipoproteins

Role of lipids

Lipids are organic molecules, largely insoluble in water, and mainly hydrocarbons of varying structure (Table 1).

Table 1
Lipids and lipoproteins

Cholesterol
Steroid molecule with hydrocarbon chain. Cholesterol ester with fatty acid is much less water soluble than free cholesterol

Triglycerides
Fatty acids of various lengths and composition are condensed onto glycerol

Phospholipids
Somewhat similar structure to triglycerides, with two fatty acids and choline phosphate on glycerol backbone

Fatty acids
Hydrocarbon chains described by length, number of double bonds and distance from the methyl group to first double bond

Triglycerides are the main fuel stores, because they can pack into smaller spaces than protein or carbohydrate, with substantially less water and twice the calories per gramme. Triglycerides are either of dietary origin or are synthesized in the liver from other 'fuels'.

Cholesterol and phospholipids are the structural components of cell membranes and organelles. Most of the cholesterol is synthesized in the liver; some is used to produce biliary sterols, but most is secreted on lipoproteins.

Structure and functions of lipoproteins

Lipids are transported in blood as large macromolecules; these are complexes with proteins. Free fatty acids are the exception, mainly binding to albumin.

Hydrophobic lipids, triglycerides and cholesterol esters are within the lipoprotein core, with the polar portions of phospholipids and the water-soluble alcohol portion of free cholesterol projecting into the aqueous environment, allowing solubilization of the lipoprotein.

Apolipoproteins have structural and functional roles: stabilization of the molecule; receptor-recognition peptides for cell lipoprotein uptake; or coenzymes for lipid metabolism.

Types of lipoproteins

Lipoprotein classes can be separated physicochemically, either by electrophoresis which uses surface charge or by ultracentrifugation which uses relative density (Table 2).

Chylomicrons

These are the largest, lightest lipoproteins (Table 2); they carry dietary triglycerides to be hydrolysed by peripheral tissue lipoprotein lipase (LPL) (Figures 5 and 6). Fatty acids either provide energy or are stored as triglycerides for later use.

Very-low-density lipoproteins (VLDL)

VLDL are the next largest lipoproteins (Table 2); they carry triglycerides synthesized in the liver also to the periphery (Figures 5 and 6).

As triglycerides are removed, chylomicrons and VLDL shrink in size, and are further catabolized as remnants. Chylomicron remnants are removed by a liver chylomicron remnant receptor, known as LRP (LDL receptor-related protein) or the α_2-macro-globulin receptor. VLDL remnants, *intermediate-density lipoproteins* (IDL), are acted on by hepatic lipase (HL), removed by the liver or converted to *low-density lipoproteins* (Table 2).

Low-density lipoproteins (LDL)

LDL are the main cholesterol carriers (Table 2), delivering cholesterol to peripheral tissues, or back to the liver, through LDL receptors (Figures 5 and 6).

Table 2
The five main lipoprotein classes and their characteristics. Between lighter, larger LDL$_2$ and smaller, denser LDL$_3$ the diameter difference is only 2–4nm

		Diameter (nm)	Density (g/ml)
	Chylomicron	100–800	≤ 0.95
	VLDL	30–80	< 1.006
	IDL	25–35	1.006–1.019
	LDL	20–25	1.019–1.063
	HDL$_2$	9–10	1.063–1.125
HDL			
	HDL$_3$	6–7	1.125–1.21

High-density lipoproteins (HDL)

HDL (Table 2) carry around 20% of plasma cholesterol in a reverse transport pathway from cholesterol-replete tissues (Figures 5 and 6).

Apolipoproteins

Apoprotein B

Apoprotein B (apoB) is the structural protein for all lipoproteins except HDL; it has two forms; $apoB_{48}$ and $apoB_{100}$ (Table 2).

Electrophoretic mobility	Main lipid in core	Main apolipoprotein	CHD risk
Origin	Dietary triglyceride	$apoB_{48}$ apoCs (apoA-I, A-II, E)	Nil (pancreatitis)
Pre-β	Endogenous triglyceride Cholesterol esters	$apoB_{100}$ apoCs apoE	+ (pancreatitis)
Broad β	Cholesterol esters Triglyceride	$apoB_{100}$ apoE	+++
β	Cholersterol esters (Triglyceride)	$apoB_{100}$	+++
α	Cholesterol esters	apoA (apoCs) (apoE)	Inverse

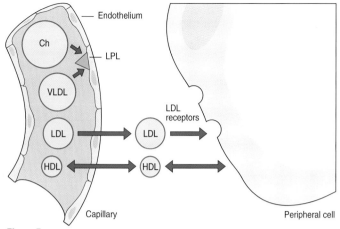

Figure 5

Simple lipoprotein function and metabolism. Dietary fat is carried on chylomicrons (Ch), and endogenous (hepatic) fat on VLDL, to peripheral tissues where lipoprotein lipase (LPL) hydrolyses the triglyceride for energy supply or storage. Chylomicron remnants are removed by the liver. VLDL remnants (IDL) are either removed by the liver or converted to LDL, the main cholesterol-carrying particles. LDL either provide cholesterol to peripheral tissues or are removed by the liver. HDL act as a reverse cholesterol transport pathway returning cholesterol from cholesterol-replete tissues to the liver or for interchange with other lipoproteins.

$ApoB_{100}$ (in VLDL, IDL and LDL) also has an LDL receptor ligand component. The mRNA for gut apoB is edited, leading to truncated $apoB_{48}$ which is structural only and is not involved in chylomicron clearance (Figure 6).

Apoprotein C

There are three C apoproteins: apoC-I, apoC-II and apoC-III; the last has three isoforms: $C-III_0$, $C-III_1$ and $C-III_2$. ApoC-II activates LPL.

Apoprotein E

Apoprotein E is involved in receptor-mediated transfer of cholesterol between the tissues and the plasma. ApoE (and

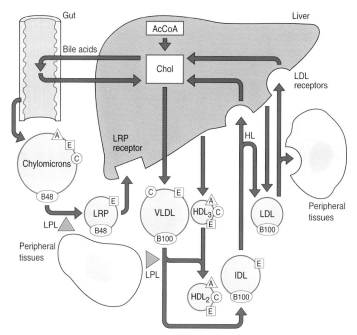

Figure 6

Lipoprotein inter-relationships and metabolism. The main apoprotein for chylomicrons, VLDL, IDL and LDL is apoprotein B. ApoB on chylomicrons is a truncated structural protein (apoB$_{48}$) compared to apoB$_{100}$ of VLDL, IDL and LDL which also has a functional, receptor-binding portion. Apoprotein Cs and E are acquired in plasma by chylomicrons and VLDL. ApoC-II activates LPL. ApoE is important for chylomicron remnant removal (through LRP or LDL receptor-related receptor protein) and IDL removal. ApoA is the main HDL protein. LCAT, associated with HDL, esterifies cholesterol to its ester. Lipid transfer between lipoproteins is facilitated by the enzyme CETP (cholesterol ester transfer protein). AcCoA, acetyl Co-enzyme A; Chol, cholesterol.

apoCs) are transferred from HDL to new chylomicrons and VLDL, transferring back during triglyceride-rich particle catabolism. Multiple apoE copies on chylomicron remnants bind to LRP receptors. ApoE binds to hepatic apoB/E receptors for removal of IDL not converted to LDL.

Apoprotein A

Apoprotein A occurs as two main proteins: apoA-I and apoA-II; HDL molecules have either one or both types. HDL of gut origin with only apoA-I are the best cholesterol acceptors. ApoA-I activates the enzyme LCAT (lecithin:cholesterol acyltransferase).

Apoprotein (a)

Apoprotein (a) is a glycoprotein which, linked to LDL, forms lipoprotein (a) (Lp(a)). Apo(a) has considerable structural similarity to plasminogen. Multiple alleles of apo(a) have various repeats that are analogous to the 'kringles' of plasminogen. The molecular size and concentration of Lp(a) are largely genetically determined, with high levels being associated with risk for CHD and probably increased thrombogenic risk.

Enzymes

Lipoprotein lipase (LPL), found in muscle and adipose tissue capillaries, hydrolyses lipoprotein triglycerides (Figure 6). *Hepatic lipase* also hydrolyses lipoprotein triglyceride and is involved in the conversion of IDL to LDL (Figure 6) and also in the formation of small dense LDL (LDL$_3$) and HDL (Figure 9).

Cholesterol synthesis is controlled by *3-hydroxy-3-methylglutaryl-coenzyme A reductase* (HMG-CoA reductase). Nonhepatic cells receive most of their cholesterol from LDL, not from endogenous synthesis. A fall in cell cholesterol up-regulates HMG-CoA reductase. A great deal of the lipoprotein-cholesterol that is secreted subsequently returns to the liver, either on LDL or from reverse transport on HDL.

Lecithin:cholesterol acyltransferase (LCAT), which accompanies HDL, esterifies free cholesterol, the ester moving to the

HDL core. HDL_3 becomes the larger HDL_2; this process may be reversed by hepatic lipase (Figure 6). LCAT aids the removal of phospholipids and free cholesterol from VLDL during catabolism, and also their passage to HDL.

Cholesterol from LDL and HDL passes to VLDL and chylomicron remnants in exchange for reverse movement of triglycerides, through the action of the enzyme cholesterol ester transfer protein (CETP) (Figure 9). This limits the accumulation of cholesterol ester in HDL acting as one reverse cholesterol transport pathway. However, excess cholesterol ester in VLDL may not be physiologically appropriate. This exchange is increased in individuals who have prolonged and post-prandial hypertriglyceridaemia (as in diabetes), and leads to abnormal and dysfunctional LDL and HDL.

Receptors

Lipoproteins are removed from the circulation by specific clearance systems.

The LDL receptor recognizes $apoB_{100}$, and IDL apoE, taking up cholesterol into the cell. Higher cholesterol levels in the cell lead to:

- increased acyl cholesterol acyltransferase (ACAT) activity with cholesterol re-esterification
- inhibition of HMG-CoA reductase and cholesterol synthesis
- decreased new LDL-receptor synthesis and membrane insertion

LRP binds multiple apoE copies on chylomicron remnants, facilitating the hepatic removal of remnants.

Atherosclerotic plaques are an inflammatory response to vessel wall injury; they contain macrophages and T lymphocytes, and have cytokine and growth factor expression. Endothelium modifies LDL, with phospholipid peroxidation and apoB degradation.

Activated macrophages with *acetyl-LDL* or *oxidized-LDL receptors* avidly remove modified LDL. Macrophage uptake may be a helpful scavenging mechanism, but, in excess, macrophages become lipid-laden foam cells of the fatty streak, which have a pivotal role in atherogenesis.[29]

Antioxidants and vitamins

Changes in diabetic metabolism that predispose to lipoprotein oxidation may accelerate atherogenesis. Antioxidants may protect LDL from oxidative damage, and may be anti-atherogenic. The antioxidant vitamin E and β-carotene are carried on LDL. Clinical trial evidence of benefit from antioxidant vitamin supplementation is incomplete at present.

In addition, folic acid may also have an anti-atherogenic role, by reducing usual blood levels of homocysteine,[30] and this may also be applicable to individuals with diabetes. Elevated blood homocysteine is associated with increased CHD risk. Results of research studies are awaited.

Diabetes may lead to hyperlipidaemia and exacerbate underlying hyperlipidaemia.

Absent LPL activity (Type I hyperlipidaemia)

Severe hypertriglyceridaemia occurs in inherited deficiency of LPL or apoC-II, when chylomicrons and VLDL are not cleared.

Clearance of VLDL remnants (Type III hyperlipidaemia; familial dysbetalipoproteinaemia)

Of the apoE isoforms, $apoE_2$ has only 1–2% of the binding ability to apoE receptors of $apoE_3$ or $apoE_4$. This affinity is sufficient to clear IDL unless there is excess VLDL synthesis or other genetic or environmental factors. Remnant lipaemia may occur in diabetes and in IGT, in overweight individuals (particularly central adiposity) or with alcohol excess. It is characterized by tubero-eruptive and palmar xanthomata (Figure 7), mixed lipaemia (cholesterol 8–12 mmol/l, triglycerides 5–20 mmol/l), and premature PVD and CHD.

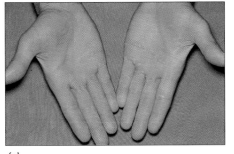

(a)

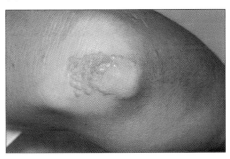

(b)

Familial hypercholesterolaemia and LDL-receptor defects

Familial hypercholesterolaemia (FH) is the result of single gene defects affecting the LDL receptor which is either defective or not produced.[31] In the heterozygote (0.2% of the population), cholesterol doubles, and quadruples in the homozygote. Half the men develop CHD by the age of 50, and half the women by 60 years of age. Some individuals have a normal receptor, but an abnormal LDL protein ('apoB3500') with impaired receptor binding.

Small dense LDL species, and oxidized or glycated LDL (more probable in diabetes), are less readily recognized by the LDL receptor.

Familial combined hyperlipidaemia

Familial combined hyperlipidaemia (FCH) has a prevalence of 1%. VLDL and apoB synthesis are increased, but the primary biochemical defect is not fully established. Cholesterol, triglycerides or both are elevated, varying between or within individuals with time, alcohol intake, lifestyle and weight. Diabetes can exacerbate FCH.

Common hyperlipidaemia

Moderate hyperlipidaemia in substantial proportions of the population reflects environment, lifestyle and disease (eg diabetes) acting on a polygenic background.

Secondary hyperlipidaemia

Diabetes commonly causes hyperlipidaemia, particularly when inadequately controlled. It may coexist with other secondary causes. Elevation of triglycerides is typical in diabetes, with low HDL in Type 2 diabetes.

Diabetes

Diabetes is arbitrarily defined by circulating blood glucose levels. Arterial, capillary and venous glucose levels will vary a little, reflecting some glucose extraction by tissues. The diagnostic criteria[32] have recently been revised,[33] and are shown in Table 3.

Patients with Type 1 diabetes lack insulin, and require insulin treatment to control symptoms and to prevent ketoacidosis and death.

In Type 2 diabetes, spontaneous ketoacidosis does not occur, individuals are often overweight and there is insulin resistance with relative insulin lack. Some Type 2 diabetes patients progress over the years to require insulin treatment.

The new American Diabetes Association criteria[33] have now been accepted by the WHO and in the UK, where the previous international criteria had been used.[32] Close to the cut-off points between normal and impaired glucose handling, and between impaired glucose handling and diabetes, substantial numbers of patients will be classified differently in the two systems. The main USA change is to use fasting and not two hours post-glucose values (leading to the term 'impaired fasting glucose' rather than IGT). The two-hour value can be used

Table 3
A diagnosis of diabetes is made by any of three criteria: (a) symptoms of diabetes and a casual (random) plasma glucose ≥ 11.1 mmol/l; (b) a fasting plasma glucose > 7 mmol/l (8 hours fast with water only); (c) a plasma glucose ≥ 11.1 mmol/l 2 hours after a 75 g oral glucose load. In the absence of unequivocal hyperglycaemia with acute metabolic decompensation, these criteria must be confirmed by a further abnormal glucose measurement on a subsequent day[32,33]

	Fasting plasma glucose	2 hour plasma glucose in 75g oral glucose tolerance test	Random plasma glucose
Normal	< 6.1 mmol/l (< 110 mg/dl)	< 7.8 mmol/l (< 140 mg/dl)	
Impaired glucose tolerance (IGT)	≥ 6.1 mmol/l and < 7.0 mmol/l (≥ 110 mg/dl to < 126 mg/dl)	≥ 7.8 mmol/l and < 11.1 mmol/l (≥ 140 mg/dl to < 200 mg/dl)	
Diabetes mellitus	≥ 7.0 mmol/l (≥ 126 mg/dl)	≥ 11.1 mmol/l (≥ 200 mg/dl)	≥ 11.1 mmol/l (≥ 200 mg/dl) with symptoms of diabetes

alone in the WHO criteria for epidemiology, but both fasting and two-hour values are needed for the individual. CHD risk is increased in 'impaired fasting glucose' patients.[5]

Patients with Type 1 diabetes lack insulin and require insulin treatment to control symptoms and to prevent ketoacidosis and death.

In Type 2 diabetes, spontaneous ketoacidosis does not occur, individuals are often overweight and there is insulin resistance with relative insulin lack. Some Type 2 diabetes patients progress over the years to requiring insulin treatment.

Metabolism, insulin and lipids

Liver glucose uptake is not insulin dependent, although insulin greatly increases glucose uptake in muscle and adipose tissue, and promotes hepatic glycogen synthesis and lipogenesis. Insulin deficiency increases glycogenolysis, gluconeogenesis, lipolysis and ketone production. Residual insulin allows ketone catabolism in Type 2 diabetes but not in Type 1 diabetes with insulin lack.

In Type 2 diabetes or partly controlled Type 1 diabetes, excess liver triglyceride and VLDL synthesis occur from fatty acids and other substrates (Table 4).

Type 2 diabetes is an hyperinsulinaemic state; higher insulin levels than in thin, non-obese, non-diabetic individuals are insufficient to compensate for insulin resistance.

Insulin may oppose glucagon inhibition of triglyceride synthesis but lack of insulin increases fatty acid levels, and therefore hepatic triglyceride, VLDL and apoB synthesis.

Table 4
Average changes in serum lipid and lipoprotein concentrations in those with Type 1 or Type 2 diabetes compared to non-diabetic control groups ('Normal levels' may still carry excess CHD risk)

	Type 2 diabetes	Type 1 diabetes
Triglycerides	↑	↑ or N
Cholesterol	N	N (or ± ↓)
VLDL triglycerides	↑	↑ or N
LDL cholesterol	N	N (or ± ↓)
HDL cholesterol	N or ↓	↑ or N

N, normal

Lipid abnormalities in diabetes

Triglycerides and VLDL triglycerides

Hypertriglyceridaemia is the most common lipid abnormality of diabetes, especially in inadequately treated or untreated patients. With good control of Type 1 diabetes, triglyceride levels are normal or slightly raised; hypertriglyceridaemia may persist in well-controlled Type 2 diabetes, reflecting factors such as overweight, β-blocker or thiazide treatment, or high alcohol intake. VLDL synthesis is increased in Type 2 diabetes and may approach double that in controls; failure to increase VLDL conversion to IDL then leads to hyperlipidaemia.[34]

Lipid and lipoprotein changes in diabetic patients, compared to matched controls, are shown in Figure 8.

Severe hypertriglyceridaemia may occur in ketoacidosis. Good diabetic control with adequate insulin treatment usually normalizes VLDL synthesis and removal. Plasma fatty acid levels may be subnormal if there is euglycaemia with normal hepatic insulin; this can only be achieved in Type 1 diabetes, by elevated systemic insulin levels.

High-density lipoproteins

HDL levels have an inverse correlation with VLDL levels. With impaired VLDL clearance, transfer of surface lipids and apoproteins from VLDL to HDL is reduced. Altered hepatic lipase activity also contributes to low HDL levels. There may be altered apoA metabolism. The HDL level is low in Type 2 diabetes, especially when associated with hypertriglyceridaemia, and may be one reason for the loss of protection in women with Type 2 diabetes. A low HDL level is compounded by insulin resistance and obesity; the central or male pattern of adiposity that predisposes to this is more common in diabetic than in non-diabetic women.

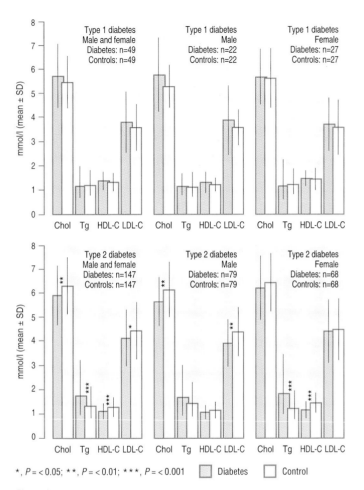

*, P = < 0.05; **, P = < 0.01; ***, P = < 0.001   ▨ Diabetes   ▢ Control

Figure 8
Plasma lipid and lipoprotein concentrations in a community-based diabetic population and their matched controls. In a geographically-defined area diabetic patients were ascertained (prevalence of diabetes 1.7%), and those aged 18–70 were approached for study. Individually age- and gender-matched non-diabetic controls were randomly obtained from the same general practices and geographical area. From Lloyd and Reckless.[35] Significant differences in patient groups for different lipids are shown. There were no significant differences between Type 1 diabetes and control groups. Females with Type 2 diabetes also had significantly higher VLDL-triglycerides (***) and higher VLDL-cholesterol values (**).

Good diabetic control will tend to normalize triglyceride levels, but HDL levels may not respond fully.

Remnant lipoproteins (IDL)

IDL normally undergo rapid conversion to LDL or are removed by the liver. An increased VLDL level is associated with slower VLDL removal and a longer period spent in the plasma. An abnormal IDL may occur; this is removed less readily by the liver, and taken up more readily by the macrophage scavenger or oxidized-LDL receptor with atherogenic potential. The VLDL-apoB proportion removed by the liver increases,[36] whereas conversion of VLDL to LDL is reduced. Consequently, the absolute production and concentration of LDL are not increased.

Overall lipid patterns in diabetes

1. On average cholesterol and LDL-cholesterol are no higher in diabetes than the non-diabetic population, although in both groups they often may be higher than desirable.

2. The classical abnormality in diabetes is a higher triglyceride level, often with too low a level of the 'protective' HDL-cholesterol.

3. Higher triglycerides lead to abnormal quality of LDL which is more atherogenic than usual, and also to abnormal HDL.

4. These changes are more common in the overweight, especially in the centrally overweight individual, and also more common in Type 2 than Type 1 diabetes.

Cholesterol and LDL cholesterol

Total and LDL cholesterol levels are usually normal (see Figure 8), unless diabetic control is poor; with intensive insulin treatment values can be low to normal.[37,38] With intensive treatment in the DCCT trial,[2] LDL levels fell by 34% ($P < 0.02$) in those patients with an LDL cholesterol level of more than 4.14 mmol/l. This represents about the top 10% of the distribution because the average LDL level was 2.90 mmol/l, and the overall LDL level did not change. Poor VLDL clearance can result in a low LDL level; correcting the VLDL defect can cause

Figure 9

Hepatic-synthesized VLDL shrink as triglycerides are hydrolysed by LPL to deliver fatty acids to peripheral tissues, and through an intermediate density lipoprotein remnant (IDL) form LDL, which in turn are removed by the liver or peripheral tissues for cholesterol provision.

Some LDL (and HDL) cholesterol ester passes to VLDL through action of CETP, with reverse passage of triglyceride (not necessarily in equimolar amounts). Some of the LDL (or HDL) triglyceride is hydrolysed by hepatic lipase activity, resulting in a slightly smaller, denser lipoprotein.

In hypertriglyceridaemic patients (such as those with diabetes), there may be increased hepatic apoB synthesis and particularly increased triglyceride synthesis, but also reduced VLDL clearance (shown in blue). As a result, increased numbers of larger VLDL particles having a longer residence time in plasma allow more CETP-mediated triglyceride–cholesterol ester exchange. In hypertriglyceridaemic diabetes there is triglyceride-enrichment of LDL (and HDL), and as a result more subsequent action of hepatic lipase (HL) leads to smaller, denser LDL particles called LDL_3 (and also to smaller HDL particles). These small, dense LDL_3 have lower affinity for, and are less readily removed by, hepatic and other LDL receptors, but are more likely to be removed by the macrophage scavenger receptor. (The width of lines and arrows reflects relative flux.)

LDL levels to rise slightly, although the quality of LDL (see below) is likely to have improved.

Changes in lipoprotein quality

Not only do total (see Figure 8) and VLDL triglyceride levels increase, but also the triglyceride content of other lipoproteins, and this may impair lipoprotein handling. Increased VLDL triglyceride synthesis is associated with larger particles, higher triglyceride:apoB ratio, slower clearance and larger remnants (Figure 9). Diabetic VLDL and LDL have a higher free cholesterol content.[39] LCAT activity may be reduced in untreated Type 1 diabetes, but not in treated Type 1 diabetes or Type 2 diabetes.

Small dense LDL

Studies of diabetic LDL subfractions showed fewer of the larger, lighter LDL_2 molecules and more of the small, dense LDL_3 molecules,[40,41] and therefore more LDL molecules for any

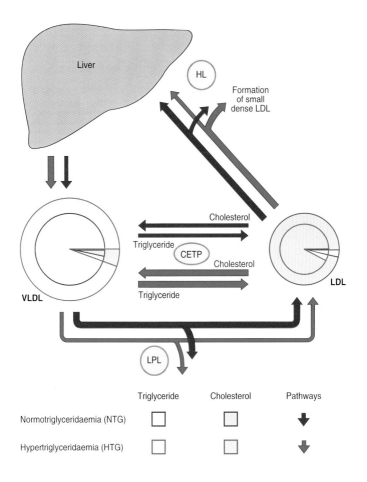

	Triglyceride	Cholesterol	Pathways
Normotriglyceridaemia (NTG)			
Hypertriglyceridaemia (HTG)			

given LDL. Slow clearance of hypertriglyceridaemic VLDL molecules allows more time for CETP transfer of cholesterol ester to VLDL and triglyceride to LDL. This LDL triglyceride is hydrolysed by hepatic lipase, giving small dense LDL_3 molecules, which have a lower affinity for the LDL receptor,[42] may undergo oxidative modification and can be removed via the oxidized LDL receptor. These diabetic changes increase the atherogenic potential of the LDL, but improved diabetic control tends to normalize the LDL subfractions,[40,41] as may hypolipidaemic therapy.[43] Statins lower LDL concentrations more effectively, but do less for lipoprotein quality than the fibrates.

Apoprotein changes

ApoB synthesis increases in poorly controlled diabetes; VLDL triglyceride synthesis is increased even more, and the IDL are then more likely to be removed directly by the liver or macrophages than be converted to LDL.

Phenotype patterns of apoE influence plasma lipids, but this influence is similar in diabetic and non-diabetic populations. ApoE$_2$ homozygosity slightly lowers LDL cholesterol levels from the slower IDL to LDL conversion. If such conversion is further reduced or VLDL are overproduced in poorly controlled diabetes, then IDL accumulate and remnant (IDL) lipaemia may occur.

Inconsistent Lp(a) changes have been reported in diabetes.

Glycation

Glycation is a major pathway for protein (including lipoprotein) modification, with non-enzymic glucose addition dependent on glucose levels, linked with glucose auto-oxidation and initiation of oxidative modification. Glycation of 2–5% of apoB lysine in diabetes reduces LDL affinity for its receptor by 5–25%.[44]

Lipoprotein oxidation

Free oxygen radicals are formed as an oxygen atom loses one electron from its normal four pairs, a replacement electron being subsequently acquired from other metabolites. Free glucose is prone to transition metal-catalysed auto-oxidation,[45] which may contribute to LDL peroxidation and be a factor in accounting for risk of CHD from glucose.[46,47]

LDL oxidation is important in atherogenesis,[44,48] inducing monocyte migration into the subintima, the transformation of monocytes to macrophages and the expression of oxidized-LDL receptors. LDL oxidation rates vary, probably partly reflecting LDL antioxidant carriage. Intake of antioxidants has been associated with less vascular disease in open studies[49,50] of self-selected and self-treated individuals. Prospective double-blind trials are needed, and antioxidant supplementation is not yet advised treatment.

Copper may catalyse LDL oxidation, whereas high iron (ferritin) levels are linked to excess CHD with hypercholesterolaemia.[51] LDL oxidation may generate LDL autoantibodies, which correlate also with carotid atheroma.[52]

Insulin resistance and syndrome X

Excess CHD accompanies a common complex termed 'syndrome X' or the 'insulin resistance syndrome' (Table 5). There is a major genetic component, which is exacerbated by lifestyle factors such as being overweight and physical inactivity. The syndrome contributes to excess CHD particularly in Asian and some other communities.

Table 5
Features of the insulin resistance syndrome

• Hyperinsulinaemia Insulin resistance	• Diabetes mellitus or IGT
	• Hyperuricaemia or gout
• Central obesity android or abdominal, as opposed to buttock and thigh (gynaecoid) obesity	• Mixed hyperlipidaemia and low HDL level (the atherogenic lipid profile)
• Hypertension	• Premature macrovascular disease
• High plasminogen activator inhibitor-1 (PAI-1) levels	• Polycystic ovarian syndrome

Gender differences

Diabetic women are more prone to macrovascular disease and lose any premenopausal sex protection, whereas HDL protection is less apparent in men than in women. The differences in lipid changes between the genders (see Figure 8) may account for some of the excess CHD risk in diabetic females.

Effects of complications of diabetes

Complications of diabetes may co-segregate, with the presence of one problem requiring a search for others. These relate to the duration and control of diabetes, but cannot be fully explained. Microvascular complications, such as renal disease, may be accompanied by macrovascular disease. About 40% of patients with Type 2 diabetes will have one or more complications already at the time of diagnosis.

Dyslipidaemia increases in severity as the protein loss progresses from microalbuminuria to nephrotic syndrome; in renal failure, low HDL levels and hypertriglyceridaemia are marked. Hypertension requires active treatment to limit renal disease progression and delays the institution of end-stage renal support. Haemostatic changes and high fibrinogen levels accompany diabetic renal disease.

Summary

Multiple qualitative and quantitative lipoprotein abnormalities contribute to diabetic atherogenic risk. Hypertriglyceridaemia is characteristic and is partly corrected by diabetes control. VLDL molecules are larger, with a higher triglyceride:apoB ratio and more cholesterol ester. Impaired clearance leads to athero-genic remnants. Excess triglyceride is also found in LDL and HDL. The HDL level is reduced in Type 2 diabetes. The LDL shift towards small dense LDL, which are more susceptible to oxidation; these are recognized by the macrophage receptor, but not so well by the LDL receptor. Lipoproteins may be gly-cated and oxidized.

Investigations and background

Investigation for hyperlipidaemia in diabetes relates to the following:

- Excess risk of CHD, CVD and PVD
- Increased risk of retinal vein and artery thrombosis
- Risk of acute pancreatitis

Individual assessment of overall macrovascular disease risk is required. New clinical guidelines on CHD prevention pay particular attention to multiple risk assessment and management, and to patients with CHD or diabetes.[53–55] Secondary causes of dyslipidaemia also need exclusion.

Lipids

In whom should they be measured?

As macrovascular disease is common, specific risk assessment should be made for every diabetic subject. Management should occur regularly from middle age, but younger patients, particularly those with other risks, should not be ignored. CHD prevalence increases with Type 1 diabetes duration and renal disease. Newly diagnosed Type 2 diabetes patients may already have considerable CHD, reflecting previous undiagnosed diabetes or impaired glucose tolerance. The absolute risk of developing a first macrovascular disease event in many diabetic subjects is as great or greater than that in non-diabetic patients who have already had an event.

Failure to treat younger patients relatively earlier at somewhat lower absolute risk thresholds, but waiting until their absolute risk rises sufficiently with age, can lead to very substantial loss of life years. Here life-time treatment benefits are important and not just the potential benefits over just 1, 5 or 10 years.

When should lipids be measured in diabetes?

A lipid profile should be measured at diagnosis, after initial sugar control. Measurements should be repeated over time depending on initial results, age and associated factors. Where hypertriglyceridaemia is present, a fasting lipid profile is essential. Diabetic dyslipidaemia is interpreted in the light of the concurrent diabetic control, the lipid hallmark of inadequate control being hypertriglyceridaemia. Primary dyslipidaemia is diagnosed when diabetic control is satisfactory.

Where should lipids be measured?

Most diabetic patients will be under the care of their primary health care team, and not under sole specialist care. Fasting samples are measured more easily at the site of primary care. Specialists involved in shared care should ensure that lipoprotein assessment has been carried out, especially in complicated patients and in those attending vascular or cardiology teams.

How often should lipids be measured?

If initial values are normal in young patients with no vascular disease or other risk factors, further measurement can be delayed 3–5 years, but patients with other risk factors, bad family history or developing renal disease should be monitored every 1–3 years. In older patients, as CHD risk increases or if lipid levels are borderline high, measurements should be more frequent.

There are natural variations in a person's cholesterol level of ± 12% (95% confidence levels) on a single measurement. As for blood pressure, long-term decisions should be made after multiple measurements. For primary prevention, drug therapy should not start until three lipid profiles are consistent over a number of weeks, after dietary control. In secondary prevention it may be best to commence statin (or fibrate) treatment at the time of the coronary or other event and monitor the response at follow-up at 3 months. Dry chemistry measurements (Reflotron, etc.) in the surgery are acceptable as an initial screen, but they have higher variability, so main laboratory

$$\text{LDL cholesterol (mmol/l)} = \text{plasma cholesterol} - \text{HDL cholesterol} - \left(\frac{\text{triglyceride}}{2.2} \right)$$

measurements should be the subsequent method.

What tests?

Fasting cholesterol, triglyceride and HDL-cholesterol levels, together with a calculated LDL level, are required. The LDL level is calculated from the Friedwald formula:

LDL calculated by the Friedwald formula is a little less reliable in diabetes, but is still used. Non-fasting samples need to be interpreted with care, but if the triglyceride levels are normal, they can be accepted for cholesterol but not for LDL.

If triglyceride levels are elevated over 4.5 mmol/l, then the HDL level as measured by some methods is unreliable and falsely

- Triglycerides > 2.3 mmol/l (or > 1.5 mmol/l*)
- LDL cholesterol > 3.5 mmol/l
- HDL cholesterol < 1.0 mmol/l
- LDL:HDL > 4

If HDL is low, the change in LDL quality from normal lighter LDL$_2$ to smaller dense LDL$_3$ occurs when triglyceride levels are above about 1.5 mmol/l.

high. With methods that can measure HDL at triglyceride levels over 4.5 mmol/l, LDL can still not be calculated. Otherwise, laboratories should not report HDL and LDL levels in this situation, although they sometimes do.

Abnormal lipid profile

A lipid profile that should cause concern is as follows:

Lipids levels and concurrent illness

Cholesterol levels fall rapidly after myocardial infarction, and must be measured within 24 hours of onset. Levels do not return to the basal level for 3 months, at which time they should be remeasured. In minor illness, measurement a month later is appropriate. Triglycerides tend to rise shortly after acute myocardial infarction. If hyperlipidaemia is found at the time of acute myocardial infarction, treatment should not be delayed for 3 months, although such early drug therapy makes drug efficacy assessment more difficult. Secondary causes should be excluded, and diet and lifestyle management instituted. A case can be made for starting lipid-lowering therapy (in addition to diet) by the time of leaving hospital after myocardial infarction, in view of the very poor uptake of lipid-lowering therapy in primary care 6 months post-infarct.[56] It is always possible to reduce lipid-lowering therapy at a later date if it is considered that the drug therapy is not needed as lipid levels are very low. Macrovascular disease requires lipid lowering, and cost-effectiveness in this group is excellent.[57]

United Kingdom Prospective Diabetes Study (UKPDS)

The UKPDS trial ran from 1977 to 1997, following 4209 patients for a mean of 10 years, and has just reported.[3,4,58–60]

Improved diabetic control significantly, and very cost-effectively, reduced 'all-diabetes complications' by 12% for an HbA_{1c} improvement of 0.9% while microvascular complications were reduced by 25% ($P<0.01$). Myocardial infarction fell by 16%, just missing statistical significance ($P=0.052$ rather than 0.01). There were no significant differences in the treatments used to lower glucose levels; rather it was the degree of glucose control that was important.

In the UKPDS hypertension substudy of 1148 patients followed for up to 10 years, a less strict (now rather lax) blood pressure control was compared to strict (now the routine) blood pressure control. Achieved blood pressures were 154/87 and 144/82, and this 10/5 difference reduced any diabetes-related endpoint by 24% and microvascular endpoints by 37%. In addition, stroke, myocardial infarct and cardiac failure were reduced by 17%, 11% and 15% respectively. There was no J-shaped curve, and the lower the blood pressure a patient had, the better. Using β-blockade or ACE-inhibition did not affect outcome, and allowing for trial size, there were no major advantages or disadvantages for each drug.

Glycaemia and blood pressure control were synergistically beneficial, and it is clear that, together with lipid control and anti-smoking measures, a global approach to diabetes management is essential (see 'Primary prevention of macrovascular disease', page 52). Overall, one might have hoped for bigger effects on CHD, as also in the glycaemic control study, and this may relate to no planned intervention on lipid levels.[3,4,58–60]

Diabetic control

Hypertriglyceridaemia is often associated with high sugar levels, so specific treatment of the dyslipidaemia should follow efforts to control the diabetes. Hypertriglyceridaemia may persist in patients who are on a diet and oral agents, but insulin therapy can ameliorate this. A low HDL level is also more likely in patients with Type 2 diabetes and this level may be improved with insulin.

Hypertension

Hypertension and dyslipidaemia in diabetes may track together; very rarely, this combination is secondary to conditions such as Cushing's syndrome, acromegaly or phaeochromocytoma. While targets for blood pressure treatment are <140/<85, those for individuals with diabetes are now stricter at <130/<80, although it is recognised that this will require multiple drug treatments in many patients.[55,58]

Renal disease and proteinuria

Proteinuria and microalbuminuria are associated with an increase in dyslipidaemia and hypertension; this results in an increased need for treatment of each condition. Treatment of hypertension slows down the deterioration in renal function. In young patients microalbuminuria probably reflects early diabetic nephropathy; in older patients it usually reflects diabetes but may reflect other renal or urinary tract disease.

Treatments

β-Blockers and thiazides in a dose-dependent manner may exacerbate dyslipidaemia. Blood pressure lowering protects renal function, but treatment with an angiotensin-converting enzyme (ACE) inhibitor may offer extra renal protection in diabetes, at least compared to calcium channel antagonists. However, in undiagnosed renal artery stenosis (which can be more common in diabetes) renal function can deteriorate acutely with the use of an ACE inhibitor, in which case the drug must be withdrawn. Renal function should be checked prior to, and 1 week and 1 month after, starting therapy. ACE inhibitors and calcium channel blockers are plasma lipid neutral, while α-blockers may improve the lipid profile. In cardiac failure, ACE inhibitors and β-blockers improve prognosis.

Concomitant diseases

Hypothyroidism, renal, hepatic and other disease should be excluded.

Multiple risk factors

Multiple risk factors for CHD increase the need to manage all factors. Every effort should be made to discourage smoking, but the persistent smoker has a higher CHD risk and benefit from treating factors such as hyperlipidaemia is therefore potentially greater.

When to refer

Diabetes care must be systematically organized in primary care, where most primary and secondary CHD prevention should occur. Diabetic individuals who develop macrovascular or microvascular problems should also be seen in hospital. In both situations adequate risk factor management is often incomplete and an organized review is essential. Hospital diabetes clinics should arrange for a review of lipids plus assessment of macrovascular disease for both new and follow-up patients. Those admitted to cardiac units, or attending vascular surgery or cardiology clinics, need assessment. With a full updated diabetes register and organized diabetes clinics, primary care management of CHD risk and lipids can easily be instituted. In both settings active audit is important, to ensure risk factor assessment, institution of appropriate treatments, and achievement of targets.

Problem areas

Treatment needs are few in those with normal lipid levels and no risk factors and clear cut in those with high lipid levels and CHD. Referral should occur if treatments do not achieve the target. Hyperlipidaemic patients with incompletely controlled Type 2 diabetes, those who need consideration for insulin, and those with hypertension, complications of diabetes or a poor family history, may need specialist referral and a period of shared care. This may also be required by those who are poorly responsive to initial lipid treatment. Provider units should discuss service provision with health authority and primary care group purchasers.

Diet, lifestyle and diabetes management

Management requires attention to all risk factors, and diabetic dyslipidaemia needs integrated approaches to diet, lifestyle, exercise, anti-smoking measures, appropriate choice of hypertension treatment and lipid-lowering drugs (Table 6). Advice must be tailored to match probable lifelong compliance.

Table 6
Strategies for the management of diabetic dyslipidaemia

Consider secondary causes
Alcohol
Myxoedema
Nephropathy

Diet and lifestyle changes
Weight loss in overweight
  individuals
Reduce fat intake
Increase exercise
Stop smoking

Improve blood glucose control
Diet
Oral agents in Type 2 diabetes
Metformin in obese Type 2
  diabetes
Optimize insulin regimen
Consider insulin in Type 2
  diabetes

Examine concurrent therapy
Thiazides
β-Blockers

Consider lipid-lowering therapy
Assess overall CHD risk
Assess life expectancy

Other CHD prevention approaches
e.g. Aspirin

Treatment principles follow British (Table 7[61] and Figure 10[55]), European[53] or New Zealand (Figure 11[54]) guidelines. A multi-factorial approach is emphasized because of:

- Increased risks for diabetic patients
- Importance of diabetic control
- Differences in lipid profile

Table 7
Simple priority groups for intervention have previously been advocated.[60] They first identify secondary prevention, and then priorities of decreasing importance. These link well to calculations of levels of absolute vascular risk derived from the New Zealand primary prevention guidelines[54]

Priority	Patient population
1 Highest	Established CHD in patients of either sex
2 High	Presence of primary genetic dyslipidaemia or other major risk factors, in either sex
3 Moderate	Asymptomatic men with no other risk factors and no CHD family history
4 Low	Asymptomatic postmenopausal women with no other CHD risk factors

Data from Betteridge et al[61]

Treatment priorities

Secondary prevention of macrovascular disease

Patients with CHD need active treatment of all risk factors for further disease, including hypolipidaemic drug treatment, providing there is a reasonable life expectancy (perhaps 3 years). Clinical CVD and PVD should be treated in the same way as these patients will also have coronary atheroma and will be at very high risk of an event even if CHD has not yet manifest clinically.

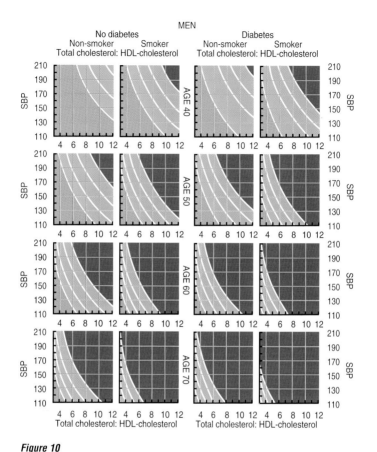

Figure 10
Joint British Societies' Charts[55] for the prediction of coronary risk in men and women. The charts are for Primary Prevention and should not be used for patients with vascular disease. For familial hypercholesterolaemia, and for individuals with renal disease, the risk will be substantially higher. SBP, systolic blood pressure (mmHg).

Identify, in turn, gender, presence of diabetes, smoking status and age to determine the appropriate chart. Within that chart, identify the intersection of systolic blood pressure and total cholesterol:HDL-cholesterol ratio. The nomogram can be used to calculate the ratio. Identify the CHD risk, given as a percentage over the next 10 years.

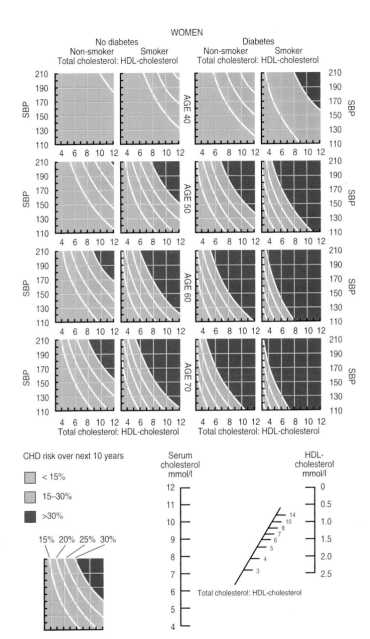

WOMEN

No diabetes / Diabetes

Non-smoker / Smoker

Total cholesterol: HDL-cholesterol

AGE 40 / AGE 50 / AGE 60 / AGE 70

SBP

CHD risk over next 10 years

< 15%
15–30%
>30%

15% 20% 25% 30%

Serum cholesterol mmol/l

HDL-cholesterol mmol/l

Total cholesterol: HDL-cholesterol

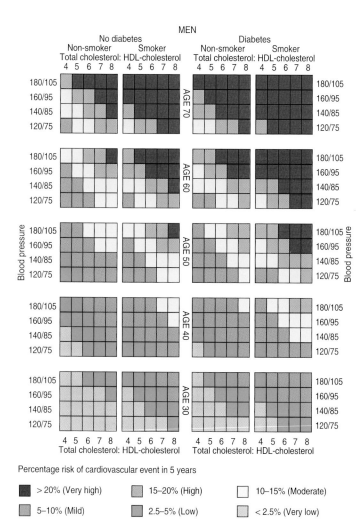

Figure 11

Estimation of cardiovascular risk for men and women for primary prevention (The New Zealand Care Service Committee[54] developed this chart from the joint European recommendations[53])

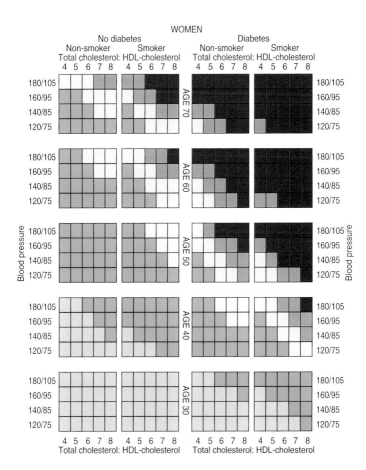

To estimate risk identify the table relating to:
(i) gender; (ii) diabetes status; (iii) smoking status; and (iv) age. Within a table
identify the cell nearest to the person's (v) blood pressure and (vi) total
cholesterol/HDL cholesterol ratio. Compare the cell colour to the key of risk levels.
It is possible to extrapolate between two adjacent blocks where necessary, for
example for age. For patients with familial hypercholesterolaemia,
hypertriglyceridaemia (when HDL low) or a bad family history of premature
vascular disease, increase their risk level by at least one colour category.

Primary prevention of macrovascular disease

The need for treatment of CHD risk factors should be based on the absolute risk of macrovascular disease. Again, notice should be taken that there is an adequate life expectancy. International guidelines suggest treatment above a 10% absolute risk of a macrovascular disease in the subsequent 5 years (about 2% per annum)[53,54] and prospectively this has been shown to be cost-effective,[62] although in the UK (prior to the latest published data) a more conservative approach at 3% per annum has been suggested to limit costs.[63–65] The recent British guidelines[55] endorse the 3% per annum as an absolute minimum, targetting individuals at $\geq 1.5\%$ per annum progressively.

For primary prevention of CHD in diabetes patients the New Zealand guidelines[54] are among the most straightforward (Figure 11). Unlike guidelines using a predetermined 3% risk level[63–65] the New Zealand guidelines allow estimation of more precise risk from $< 0.5\%$ to $> 4\%$ per annum. The New Zealand guidelines should be applied to primary prevention patients, and the calculated absolute risk should be increased by one (or more) categories if there are additional risks such as a bad family macrovascular disease history, or if there is hypertriglyceridaemia associated with a lower HDL level.

All risk factors should be addressed, and not singly considered. Where risk is sufficient hypotensive and/or hypolipidaemic therapy should be considered.

The four British societies (Cardiac Society, Hyperlipidaemia Association, Hypertension Society and Diabetic Association) have agreed guidelines for heart disease prevention.[55] For patients with macrovascular disease and diabetes, cholesterol and LDL will need to be lowered towards 4, and around 2.5, mmol/l respectively. For primary prevention, the four British societies[55] have used the Framingham equations to calculate absolute risk, as have the New Zealand guidelines. Unlike the New Zealand charts, both blood pressure and plasma cholesterol/HDL cholesterol ratio are considered as continuous variables (Figure 10). The absolute risk is calculated over a ten year period (rather than five years as in the New Zealand charts), both giving different levels of risk. A risk of 1.5–2% per annum (15–20%/10 yr British; 7.5–10%/5 yr New Zealand) normally justifies drug intervention. Both consider diabetes, and either chart can be used. The British chart is probably more accurate but the coloured boxes of New Zealand may be easier to use. Guidelines are only approximate and are not a precise prediction for any individual patient. In both sets of guidelines, family history should influence decision making while high triglycerides with low HDL would indicate an increased risk and so a need for treatment.

National Service Framework and CHD prevention

In the UK a National Health Service initiative for clear quality standards has led to a National Service Framework for CHD.[66] It targets both improved intervention rates and more prevention targeted by risk. They endorse active secondary prevention, and also primary prevention at 1.5–3% risk per annum, the lower level as resource allows. Cholesterol should fall by >30% and also to <5 mmol/l, and LDL cholestorol should fall to <3.0 mmol/l. Treatment is clearly cost-effective at an annual risk of ≤1.5% per annum,[67] and almost all dyslipidaemic patients with diabetes have risks substabtially above this. A National Service Framework for Diabetes, currently in preparation, will further address these issues in the coming year.

Diet

Diet focuses on diabetic control, sugar limitation, increase in fibre and complex carbohydrate intake, weight loss in obese individuals and reduced fat intake (Table 8).

Dietary details

Restricting sugar requires an increase in calories from other sources. Fat is inappropriate as a source of energy, and complex carbohydrates and food rich in fibre (especially soluble) can improve diabetic control.[68,69]

Reduced fat intake will reduce plasma and LDL cholesterol levels. Short-term, high-carbohydrate diets in diabetes increase VLDL, although not long term. Substituting complex carbohydrates for saturated fat can improve glucose tolerance and insulin secretion, and can lower LDL levels by 10%.[70] Plant sterols can also lower cholesterol, and are marketed in the fat spread Benecol™.

Table 8
Principles of dietary management of diabetic dyslipidaemia

---

1 **To improve diabetic control:**
   (a) reduction of refined carbohydrate
   (b) increase in fibre and complex carbohydrate

2 **To reduce weight (where required):**
   (a) reduction in total calorie intake
   (b) reduction in total fat intake
   (c) reduction in refined carbohydrate
   (d) reduction in alcohol intake
   (e) increase in fibre and complex carbohydrate

3 **To manage the dyslipidaemia:**
   (a) reduction of saturated fat intake
   (b) reduction of dietary cholesterol
   (c) partial replacement with monounsaturated and polyunsaturated fats
   (d) reduction in alcohol intake (especially in individuals with high triglyceride levels)

4 **To manage associated hypertension:**
   (a) reduction of salt intake
   (b) reduction of alcohol

Recommended foods[53] (Table 9) are generally low in fat and/or high in fibre. Foods 'to avoid' contain substantial saturated fat or cholesterol. Foods to 'use moderately' containing unsaturated or some saturated fats, include:

- Lean red meat – up to three times weekly
- Medium fat cheeses, meat and fish pastes – weekly
- Homemade cakes, biscuits, pastries (mono- or polyunsaturated oil or margarine) – twice weekly
- Chips or roast potatoes (cooked in suitable oil) – once every two weeks.

For overweight individuals additional requirements are:

- Portion control
- Alcohol avoidance
- Daily exercise

Diet and diabetic dyslipidaemia

Non-diabetic and diabetic diets are similar, but refined sugar restriction makes the diet for diabetes more demanding.

LDL unsaturated fatty acids may undergo peroxidation, but they increase hepatic LDL removal in comparison to saturated fat. However, peroxidation is less, and lipid levels tend to fall, with mono- and polyunsaturated fats compared to saturated fats;[71] this may be because the half-life of these LDL is shorter.

Fibre

Stone and Connor[72] showed falls in cholesterol level of 20% and in triglyceride level of 15% when high complex carbohydrates were substituted for restricted carbohydrates, and many other studies have been carried out. Inclusion of soluble fibre improves lipid and glucose control.[73]

Table 9

Recommended foods in a lipid-lowering diet. While suitable in a lipid-lowering diet, some acceptable foods may need limitation from a diabetic standpoint
Data from Pyorala et al[53]

Recommended foods	Foods for use in moderation	Foods to be avoided
Cereals Wholegrain bread Wholegrain breakfast cereals Porridge Cereals Pasta Crispbread Rice		Croissant Brioche
Dairy products Skimmed milk Very-low-fat cheeses, eg cottage cheese, fat-free fromage frais or Quark Very-low-fat yoghurt Egg white Egg substitutes	Semi-skimmed milk Fat-reduced and lower fat cheeses eg Brie, Camembert, Edam, Gouda, Feta, Ricotta Low-fat yoghurt Two whole eggs per week	Whole milk Condensed milk Cream Imitation milk Full-fat cheeses Full-fat yoghurt
Soups Consommés Vegetable soups		Thickened soups Cream soups
Fish All white and oily fish (grilled, poached, smoked). *Avoid skin (eg on sardines or whitebait).*	Fish fried in suitable oils	Roe Fish fried in unknown or unsuitable oils or fats
Shellfish Oysters Scallops	Mussels Lobster Scampi	Prawns Shrimps Squid
Meat Turkey Chicken Veal Game Rabbit Spring lamb	Very lean beef Ham Bacon Lamb (once or twice a week) Veal or chicken sausage Liver twice a month	Duck Goose All visibly fatty meats Usual sausages Salami Meat pies Pâtés Poultry skin
Baked foods	Pastry Biscuits prepared with unsaturated margarine or oils	Commercial pastry Biscuits Commercial pies Snacks and puddings

Recommended foods	Foods for use in moderation	Foods to be avoided
Fats	Polyunsaturated oils eg sunflower, corn, walnut, safflower	Butter
		Suet
		Lard
	Monounsaturated oils eg olive oil	Dripping
		Palm oil
	Soft (unhydrogenated) margarines based on these oils, especially low-fat spreads	Hard margarines
		Hydrogenated fats
Fruit and vegetables		
All fresh and frozen vegetables, emphasis on legumes: beans, dried beans, lentils, chick peas	Roast or chipped potatoes cooked in permitted oils	Roast or chipped potatoes
		Vegetables or rice fried in unknown or unsuitable oils or fats
Sweetcorn		Potato crisps
Boiled or jacket potatoes		Oven chips
All fresh or dried fruit		Salted tinned vegetables
Tinned fruit (unsweetened)		
Desserts		
Sorbet		Ice cream
Jellies		Puddings
Puddings based on skimmed milk		Dumplings
Fruit salad		Sauces based on cream or butter
Meringue		
Confectionery		
Turkish delight	Marzipan	Chocolate
Nougat	Halva	Toffees
Boiled sweets		Fudge
		Coconut bars
		Butterscotch
Nuts		
Walnuts	Brazils	Cashews
Almonds	Peanuts	Coconut
Chestnuts	Pistachios	Salted nuts
Beverages		
Tea	Alcohol	Chocolate drinks
Filter or instant coffee	Low-fat chocolate drinks	Irish coffee
Water		Full-fat malted drinks
Calorie-free soft drinks		Boiled coffee
Dressings, flavourings		
Pepper	Low-fat salad dressings	Added salt
Mustard		Salad dressings
Herbs		Salad cream
Spices		Mayonnaise

Obesity

Obesity (especially central) is important in diabetic dyslipidaemia and is associated with insulin resistance, with calorie restriction and physical exercise limiting both. Individuals who carry out manual work tend to be thinner, more insulin sensitive, less hyperlipidaemic and less glucose intolerant. Populations in which lifestyle has changed (such as Pacific Island communities) have more obesity, hyperlipidaemia and diabetes in urban, compared to rural, dwellers.

Smoking

To succeed in giving up smoking, a smoker must wish to stop. Smoking habits should be recorded and reviewed at follow-up, and continued support to quit should be provided, especially for those expressing such a wish. Explicit reasons and advice should be given, literature provided and counselling given. Nicotine transdermally, by inhalation, or in chewing gum, may be useful. Those who quit should avoid situations where there are other smokers.

Alcohol

Alcohol intake has CHD risk benefits, but it should be limited to ≤ 14 units/week for women and ≤ 20 for men. Alcohol is high in calories, and should be limited in overweight individuals. As hypertriglyceridaemia may be exacerbated by alcohol, alcohol should be limited in such individuals, especially where overweight is also present. The amount of alcohol drunk may not be volunteered by the patient, but γ-glutamyl transpeptidase and red cell volume can give clues. Alcohol inhibits hepatic gluconeogenesis; this makes Type 1 diabetes patients more prone to hypoglycaemia, while reducing its recognition.

Exercise

Exercise exerts acute insulin-like effects by increasing membrane GLUT4 glucose transporters resulting in improved insulin sensitivity. Exercise is less effective in the thin insulin-deficient diabetic individual than in insulin-resistant overweight diabetic individuals. Daily vigorous exercise reduces insulin response to glucose, by a mechanism of decreased insulin secretion balanced by increased insulin action. Regular exercise should also limit adiposity.

Drugs

Sulphonylureas

Low HDL levels and hypertriglyceridaemia occur in Type 2 diabetes, especially in sulphonylurea-treated individuals; this may reflect metabolic severity and poor diabetic control leading to sulphonylurea treatment rather than being secondary to the drug itself. Modest weight gain is undesirable with sulphonylureas, but can occur when calorie loss through glycosuria is reduced. In the UKPDS, a mean weight loss of 4.5 kg occurred with diet prior to randomization, followed by a mean weight gain over 10 years of 2.5 kg in the conventionally treated (less strict) group compared to 4.5 kg in the more tightly controlled group. There were no significant differences between insulin treatment or the two sulphonylureas, but weight gain was less with metformin.[3,4,58–60] Changes in composition and distribution of Type 2 diabetes lipoproteins improve with better control,[41] although low HDL levels may persist.

Metformin

Metformin, associated with some weight loss, is more effective in the obese patient; it improves peripheral tissue insulin sensitivity. Combination with a sulphonylurea is appropriate. Unfortunately, gastrointestinal side effects may limit metformin use.

In the UKPDS, metformin used as first agent in obese patients was very effective, but when used in combination with other drugs was potentially detrimental. Numbers in both groups were small, and overall metformin had a moderate, non-significant benefit.[3,4,58–60]

Acarbose

Acarbose inhibits small bowel α-glucosidase activity, delays starch digestion and glucose absorption, and may improve glycaemic control and lipid levels. Flatus and lower gastrointestinal side effects may also limit usefulness. In a limited substudy in the UKPDS, those who took acarbose obtained some improvement in glycaemic control.

Thiazolidinediones

These agents are insulin sensitizers working through nuclear receptors, the peroxisome proliferator activated receptor-gamma (PPARγ), to increase hepatic β-oxidation, to reduce synthesis and increase clearance of VLDL, and to improve lipid profiles. They reduce requirement for oral hypoglycaemic agents or insulin in Type 2 diabetic patients. The first of these, troglitazone, previously available in Japan and the USA, was effective and well tolerated, but associated with occasional severe liver disease. Since the introduction in the USA of rosiglitazone and pioglitazone with better safety profiles, troglitazone has been withdrawn. These new 'glitazones' should enter the European market in 2000.

β-Cell stimulants

A new group of drugs, of which repaglinide is the first example, are insulin secretagogues through a receptor separate from the classical sulphonylurea receptor. They function best post-prandially and are less likely to give hypoglycaemia in the fasting state.

Gut lipase inhibitors

Orlistat is a newly introduced inhibitor of fat digestion (inhibiting gut lipase) which can achieve 5–6 kg more weight loss in 1 year than seen with diet alone. Side effects relate to the fat malabsorption and is fat-intake and drug-dose related. It may come to have some value in overweight Type 2 diabetic patients. The expected reduction in blood pressure, and improvement in glucose tolerance is seen, while the fall in lipids is in excess of that expected for the degree of weight loss, probably linked to its mode of action in limiting fat absorption.[74]

Insulin

Optimizing insulin improves glycaemia, and quantity and quality of lipoproteins,[40] although excess weight gain should be avoided. Abnormal lipoprotein quality may limit the reliance placed on the protective effects of HDL in diabetes.[24] Significant dyslipidaemia should encourage insulin initiation in poorly controlled Type 2 diabetes patients in place of full doses of oral agents.

Occasional patients require insulin because of severe dyslipidaemia, independent of glycaemic control. These are usually overweight patients, in whom weight has not gone down, other treatments have failed, and pancreatitis or other complications may or may not have occurred. Underlying genetic dyslipidaemia exacerbated by glucose intolerance is usual, but in some people glycaemia control will be good. Patients are insulin resistant and rarely become hypoglycaemic. Insulin may reduce triglyceride levels from 20 mmol/l or more to 3–7 mmol/l, but rarely to normal, and lipid-lowering drugs may also be required. In the UKPDS,[3,4,58–60] insulin therapy did not have increased risk compared to oral hypoglycaemic agents in Type 2 diabetic patients, and in particular there was no evidence of significantly increased CHD risk.

New insulin analogues produced by insulin gene modification are increasingly becoming available, and may allow insulin treatments better targeted to excercise and meal profiles to reduce glucose excursions. Oral and inhaled insulin analogues are being developed. If these are successful, it is uncertain whether lipid profiles will improve substantially. Insulin delivery into the portal circulation (such as with islet cell transplantation) may achieve this by obtaining hepatic insulin concentrations without systemic hyperinsulinaemia.

Hypolipidaemic agents

Where diabetic dyslipidaemia does not respond to diet and lifestyle, lipid-lowering drugs should be considered. Global CHD risk should be assessed, as the excess risk in diabetes makes lipid lowering more appropriate. No major CHD intervention trial has yet been completed in diabetes, but some are in progress. Some information on diabetes sub-groups is available, but therapy is partly based on extrapolation. Intervention levels and therapeutic target levels have been proposed[75] (Table 10). These are too conservative in the higher risk diabetic patients, where CHD risk is as high as in many secondary prevention populations, and where targets should be at least the same as in the non-diabetic population. Here, cholesterol values towards 4 mmol/l, LDL cholesterol around 2.5 mmol/l and triglycerides below 1.5 mmol/l would be desired. Many diabetics need to be managed as if they already had vascular disease.

Statins are effective where hypercholesterolaemia is the main abnormality, but fibrates are often required where there is mixed hyperlipidaemia, alone or with a statin. Such statin–fibrate combinations are effective and well tolerated.[23] General indications and side effects remain the same, but some issues are specific to diabetes.

Table 10
Targets for lipid levels in Type 2 diabetes
Therapeutic targets set in 1988 for lipid levels in Type 2 diabetes*

	Serum cholesterol (mmol/l)	Fasting serum triglycerides (mmol/l)	Serum HDL cholesterol (mmol/l)
Good control	< 5.2	< 1.7†	> 1.1
Adequate control	< 6.5	< 2.2	> 0.9
Poor control	≥ 6.5	≥ 2.2	≤ 0.9

These are now too conservative in higher risk individuals
†*The abnormal increase in LDL_3 as a proportion of total LDL begins to occur at triglyceride concentrations above 1.5 mmol/l*
Data from Alberti and Gries[75]

Current targets for lipid levels in Type 2 diabetes

Lipid targets	Cholesterol	Triglycerides	LDL cholesterol
	< 5 mmol/l	≤ 1.5 mmol/l	≤ 2.5 mmol/l

Targets have changed a little as new evidence is available[76]
HDL elevation is desirable but will follow other treatment interventions.
Low HDL levels should be considered as an additional risk factor
Rather than 'poor, adequate or good' criteria, intervention follows when risk is high, endeavouring to treat to target

CHD prevention and recent randomized controlled trials

In recent years results of large primary and secondary CHD prevention studies have reinforced the need for active risk factor intervention.

In the 4S (Scandinavian Simvastatin Survival Study) trial 4444 men and women (20%), aged 35–70, with initial cholesterol of 5.5–8.0 mmol/l, received (post-myocardial infarct) a modal dose of 26 mg simvastatin or placebo for 5.4 years. CHD mortality, CHD events, coronary artery bypass surgery and total mortality fell by 42, 34, 37 and 30% respectively, and side effects were the same in both groups.[77]

In the USA CARE (Cholesterol And Recurrent Events) study 4159 men and women (14%), aged 21–75, with initial cholesterol < 6.2 mmol/l received (post-infarct) 40 mg pravastatin or placebo for 5 years. Coronary deaths and events fell 24%, and side effects were greater in the placebo group. Mean LDL-cholesterol fell from 3.6 to 2.6 mmol/l, but the 23% of the group whose initial cholesterol was ≤ 3.2 mmol/l, while not benefitting, were not harmed.[78] Furthermore, in this and other studies, prospective reductions in stroke rates from cholesterol lowering have been demonstrated.

The LIPID study, where 9154 post-infarct patients were given 40 mg pravastatin or placebo for 5 years, was stopped because of early benefit.[79] CHD death fell by 24% ($P = 0.004$) and fatal and non-fatal infarcts by 29% ($P < 0.0001$). While there was uncertainty in the CARE study about benefit of treatment when pre-treatment LDL-cholesterol was ≤ 3.2 mmol/l, in the LIPID study there was benefit in the lower tercile for LDL-cholesterol but this was less than in the other two terciles at 16% CHD reduction, and just failed to reach significance. The small sub-group of patients with diabetes behaved similarly to the whole group.

In primary prevention, the West of Scotland (WOSCOPS) trial treated 6595 men, aged 45–65 years and with initial cholesterol 6.4–7.6 mmol/l, with 40 mg pravastatin or placebo for 4.9 years. CHD death and non-fatal infarcts fell 29% ($P < 0.001$) and total mortality by 22% ($P = 0.051$).[80]

In the AFCAPS/TEXCAPS (Air Force and Texas Coronary Atheroma Prevention Study) a more healthy lower risk group of 5608 men and 997 women were treated for 5.2 years with 20–40 mg lovastatin or placebo. This was a primary prevention study in a lower risk group (5.6% 5-year CHD risk) albeit chosen for a low HDL cholesterol of 0.96 mmol/l. There was a 37% reduction in first acute major coronary events, from an LDL cholesterol reduction of 25% to 2.96 mmol/l and a 6% rise in HDL cholesterol.[81]

The Medical Research Council – British Heart Foundation-sponsored Heart Protection Study should report by 2001, with 20 000 higher risk individuals with lower or normal cholesterol levels being treated for 5 years with 40 mg simvastatin or placebo, and has approaching 6000 individuals with diabetes.

Fibrates

Fibrates increase LPL levels, reduce hepatic apoB and VLDL synthesis, and increase LDL-receptor activity, probably by modifying a peroxisome proliferator-activated nuclear receptor (PPARα). Gemfibrozil reduced CHD in non-diabetic individuals, particularly with hypertriglyceridaemia and low HDL levels in the Helsinki Heart Study.[21] Of 4081 men in the study, 135 with Type 2 diabetes had twice the CHD rates of controls (7.4% compared to 3.3%). The diabetic patients treated with gemfibrozil had one-third the CHD rate of placebo-treated diabetic patients (3.4% compared to 10.5%).[22] Recently, the VA-HIT study showed a 22% CHD reduction with gemfibrozil treatment in a male population with lower HDL cholesterol values. The diabetic subgroup had a similar and significant CHD reduction of 25%.[23] Further large fibrate studies, with diabetic populations, are needed and awaited. The Bezafibrate Israeli Prevention Study in a non-diabetic population has not yet been published but is reported as not showing benefit, except in the sub-group with higher baseline triglycerides.

Fibrates lower triglyceride levels, whereas cholesterol lowering varies, with gemfibrozil having less of an effect on LDL levels than ciprofibrate and fenofibrate. Fibrates lower total LDL concentrations, but also substantially improve the LDL subfraction profile, reducing the small dense LDL_3 fraction from around 40–45% that can be seen in moderately hypertriglyceridaemic Type 2 diabetic patients towards a more usual figure of about 25%.

The fibrinogen level is raised in many diabetic individuals, contributing to CHD risk, but this is reduced by bezafibrate, ciprofibrate and fenofibrate. Glycaemia can improve and levels of uric acid may fall. When there is marked proteinuria and marked renal impairment, fibrates should be reduced in dose or not used.

The cost of drugs related to their cholesterol-lowering efficacy varies[82] (Table 11). Calculations do not consider HDL, triglyceride or fibrinogen levels, and should not be interpreted too strictly. Lipid responses to drug therapy can also vary considerably between individuals.

Cholesterol transport and HDL

The macrophage takes up LDL and particularly altered or oxidized LDL through the SR-A or CD36 receptor.

Cells that are cholesterol replete express an ABC1 receptor to which binds nascent HDL. This leads to hydrolysis of cholesterol ester, transport of free cholesterol to the cell surface and efflux on to the HDL. Through the HDL-associated enzyme ACAT the cholesterol is esterified and enters the HDL core as cholesterol ester.

Different fates for this HDL–cholesterol ester can occur. Firstly, cholesterol ester can be passed to VLDL and LDL through

Table 11

Approximate efficacy lipid lowering (and cost per 10% reduction) of different hypolipidaemic drugs and doses

(a) Statins

Drug	Dose	Cost p.a.	Total Chol % fall	Cost £ p.a./ 10% chol	LDL-C % fall	Cost £p.a./10% LDL fall	Triglyceride % fall	HDL % rise
Atorvastatin	10mg	246.11	29	84.87	39	63.11	19	+6
	20mg	398.89	33	120.88	43	92.77	26	+9
	40mg	613.20	37	165.73	50	122.64	29	+6
	80mg	1226.40	45	272.53	60	204.40	37	+5
Cerivastatin	100µg	168.81	16	105.51	22	76.73	9	+5
	200µg	226.17	18	125.65	25	90.47	16	+9
	300µg	237.25	22	107.84	31	76.53	16	+8
	400µg	226.17	24	94.24	34	66.52	16	+9
Fluvastatin	20mg	165.81	17	97.54	22	75.37	12	+3
	40mg	165.81	19	87.27	25	66.32	14	+4
	80mg	331.62	27	122.82	36	92.12	18	+6
Pravastatin	10mg	210.92	16	131.83	22	95.87	15	+7
	20mg	387.03	24	161.26	32	120.95	11	+2
	40mg	387.03	25	154.81	34	113.83	24	+12
Simvastatin	10mg	235.03	23	102.19	30	78.34	15	+12
	20mg	387.03	28	138.23	38	101.85	19	+8
	40mg	387.03	31	124.85	41	94.40	18	+9
	80mg	387.03	36	107.51	47	82.35	24	+8

Note: Values for changes in individual lipids are from published literature, but have some inconsistencies and should be considered approximate. They will vary depending on the patient group and pre-treatment lipid levels.

National treatment targets for higher-risk hypercholesterolaemic patients are cholesterol <5mmol/l, LDL-cholesterol <3mmol/l, or 30% reduction, whichever is greater.

Table 11 (cont.)

(b) Fibrates & Resins

Drug	Dose	Cost p.a.	Total Chol % fall	Cost £ p.a./ 10% chol	LDL-C % fall	Cost £p.a./10% LDL fall	Triglyceride % fall	HDL % rise
Bezafibrate Mono	400mg	105.85	12	88.21	16	66.16	50	+50
Ciprofibrate	100mg	174.42	22	79.28	16	109.01	–	–
Fenofibrate Micro	67mg	128.76	12	107.30	16	177.21	50	+25
	200mg	283.53	25	113.41				
	267mg	283.53	27	105.01				
Gemfibrozil	1200mg	194.88	11	171.68	8	–	50	+12
Cholesty -ramine	3 sachets	384.35	18	213.53	22	174.70	–	–
	6 sachets	768.70	30	256.23	38	202.29	–	–
Colestipol	3 sachets	437.27	18	242.93	22	198.76	–	–
	6 sachets	874.54	30	291.51	38	230.14	–	–

Note: Values for changes in individual lipids are from published literature, but have some inconsistencies and should be considered approximate. They will vary depending on the patient group and pre-treatment lipid levels.
LDL falls tend to be smaller in mixed lipaemia, where fibrates may be more effective in lowering triglycerides, raising HDL and improving LDL quality.
National treatment targets for higher-risk hypercholesterolaemic patients are cholesterol <5mmol/l, LDL-cholesterol <3mmol/l, or 30% reduction, whichever is greater.

cholesterol ester transfer protein (CETP), with some exchange in the opposite direction of triglyceride. Secondly, the HDL particle can be removed and degraded by the liver. Thirdly, the cholesterol carried on HDL (and on LDL) can be off-loaded from the lipoprotein by the liver receptor SRB1 and internalized for hepatic metabolism. Thus, most cholesterol can return to the liver or other lipopoteins without HDL catabolism.

Statins

Statins inhibit HMG-CoA reductase, which is the rate-limiting step in cholesterol synthesis. LDL-receptor activity increases and LDL levels fall by 25–50%, with moderate reduction in triglyceride and increase in HDL levels. Atorvastatin, cerivastatin, fluvastatin, pravastatin and simvastatin are available in the UK, but lovastatin is not. Major outcome data from randomized clinical trials are available for simvastatin, pravastatin and lovastatin.

Statins are well tolerated, have few side effects and can be used with renal impairment; rarely, they can cause marked rise in liver transaminases or an acute myositis with raised creatinine kinase, first seen in patients co-treated with cyclosporin A. The risk of acute myositis with statins (or fibrates) is about 1 in 30–100 000 patient-years of treatment. Such side effects are more common when fibrates and statins are combined; such a combination is outside drug data sheet recommendations, but is increasingly being used, usually in specialist care, in higher risk patients who have not responded to a single agent. For example, satisfactory use of a fluvastatin with fenofibrate combination has been reported.[83] Use of such combinations can be valuable for management of resistant diabetic dyslipidaemia.

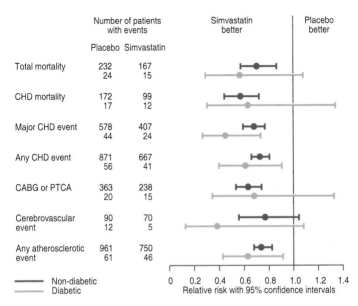

	Number of patients with events		Simvastatin better	Placebo better
	Placebo	Simvastatin		
Total mortality	232	167		
	24	15		
CHD mortality	172	99		
	17	12		
Major CHD event	578	407		
	44	24		
Any CHD event	871	667		
	56	41		
CABG or PTCA	363	238		
	20	15		
Cerebrovascular event	90	70		
	12	5		
Any atherosclerotic event	961	750		
	61	46		

——— Non-diabetic
——— Diabetic

0 0.2 0.4 0.6 0.8 1.0 1.2 1.4
Relative risk with 95% confidence intervals

Figure 12
Reduction in the risk of different endpoints expressed as relative risk (simvastatin group versus placebo group) with 95% confidence intervals in non-diabetic and diabetic patients.[77,84]

The 4S study included 202 individuals with diabetes, 105 of whom were treated with simvastatin. The diabetic patients will not have been entirely representative of general diabetes but show clear benefit.[84] The relative risks were reduced in all groups treated with simvastatin compared to placebo, and diabetic patients did as well as or better than the non-diabetic population (Figure 12 shows the relative risks and 95% confidence intervals), with the number of individuals remaining free of vascular disease after 5–6 years of follow up being improved by simvastatin, especially in the diabetic population (Figure 13).

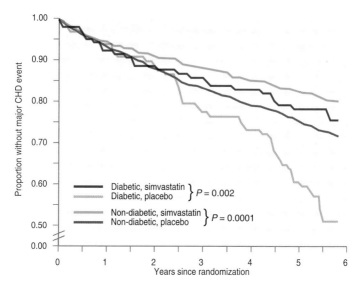

Figure 13
Kaplan–Meier survival curves for the probability of remaining free of a major CHD event during follow-up in non-diabetic and diabetic patients treated with placebo or simvastatin in the 4S trial.[84]

In the CARE population, the diabetic subgroup showed the same level of benefits as the non-diabetic group.[78] In the LIPID study the same is true.[79] Simvastatin was more effective in those diabetic patients with higher triglycerides.[85] When the new American Diabetes Association criteria for the diagnosis of diabetes[33] are applied in an analysis of the 4S trial patients' data, a further 281 patients were classified as having diabetes (fasting plasma glucose ≥ 7 mmol/l) and another 678 had impaired fasting glucose (6.0–6.9 mmol/l). In these groups, CHD events, deaths and revascularizations were reduced by 40–56% by simvastatin.

For secondary prevention, it is now clear cut that the diabetic population requires active treatment. Individuals (diabetic and non-diabetic) with PVD and CVD are unlikely to have unaffected coronary arteries, and should be treated. Indeed placebo treatments over years are no longer acceptable, and this has led to one study (the UK CARDS study) closing its smaller secondary prevention arm, but continuing recruitment to a larger primary prevention arm for normocholesterolaemic subjects with diabetes, examining the potent atorvastatin compared to placebo. Primary prevention data are limited but nothing suggests that diabetic patients are materially different, and after assessment of global risk factors, absolute risk level and evidence for adequate potential longevity, should be treated on the available general evidence. Indeed as diabetic patients have CHD risk increased two- to three-fold in males and three- to five-fold in females, their risk is such that most should be treated actively.

In the Heart Protection Study where nearly 6000 diabetic patients were prospectively recruited for a primary endpoint, the potential statin benefits in diabetes should be clearly confirmed.

In studies published to date, some inclusion criteria limitations were imposed, limiting inclusion of Type 1 diabetes patients and of Type 2 diabetes patients with more marked hypertriglyceridaemia. In the latter group, fibrate studies are needed, or combination studies using fibrate plus statin. The UKLDS (Lipides in Diabetes Study) 5 year diabetes study of cerivastatin, fenofibrate, combination or placebo has just commenced in patients with relatively normal lipid levels.

Bile acid sequestrant resins

Cholestyramine and colestipol bind gut bile acids, are not absorbed and are excreted in the faeces. They limit bile acid resorption, increase bile acid synthesis and increase hepatic LDL-receptor activity. They are rarely the first line of treatment in diabetic dyslipidaemia, but are a useful addition to statin or fibrate. Gastrointestinal side effects limit compliance, but these resins are safe and prevent CHD.[86]

Nicotinic acid and derivatives

There are good reasons why, in theory, nicotinic acid should be a lipid drug of choice, but this is not borne out in practice. Nicotinic acid limits adipocyte lipolysis and fatty acid flux to the liver for triglyceride synthesis. Regression studies on lipid-lowering, atheroma regression and CHD reduction with nicotinic acid are impressive. Nicotinic acid should improve glucose tolerance but, in practice, the reverse is seen perhaps because of fatty acid post-dose rebound. Delayed release preparations do not overcome this, and may increase disturbances of liver function. Practical problems relate to flushing, gastrointestinal disturbance and diarrhoea. It is necessary to increase the dose gradually from a low one up to 2–4 g/day.

Analogues, eg tetranicotinoylfructose and acipimox, are tolerated better, can be used with fibrates and may improve glucose tolerance, but they are less effective.

Fish oils

Fatty acids of the ω-3 series can reduce hypertriglyceridaemia, although deterioration of glucose tolerance can occur. Current preparations (Maxepa) provide 90 cal/day, which may not be helpful in the overweight, hypertriglyceridaemic diabetic patient. Fish oils may have benefits for thrombosis and coagulation, but large, controlled, long-term studies are lacking.

Antioxidants

Antioxidants might prevent CHD, but currently there is insufficient evidence to advocate widespread use. In addition to the simvastatin arm, the Heart Protection Study of 20 000 subjects has an antioxidant vitamin arm (vitamin E, vitamin C, β-carotene), and will provide further evidence, in general and diabetic populations, and by gender.

Garlic

Garlic has moderate beneficial effects on lipid profiles and this is seen in Type 2 diabetes patients.[87]

Other approaches

A future non-absorbed gut ACAT inhibitor might limit fat absorption and hyperlipidaemia. However, an absorbed ACAT inhibitor may have unpredictable effects.

Orlistat is a new non-absorbed inhibitor of gut lipase, which, when taken as part of a calorie-controlled, lower-fat, weight-reducing diet, will lower blood pressure, improve insulin sensitivity and lower glucose, and will lower lipids. Indeed, because the mode of action leads to loss of up to 30% of fat undigested in the stool, the effects on lipids may be greater than weight loss itself might suggest. The stool fat loss may of course give rise to lower gastrointestinal side effects.

As lipoprotein quality is abnormal in diabetes, with the interchange of lipids among lipoproteins, and as increased expression of CETP appears atherogenic, future development of a CETP inhibitor may have therapeutic potential.

Further developments in modification of nuclear receptor activity modulating gene expression for lipid and lipoprotein metabolism may be of benefit, as with the thiazolidinediones as insulin sensitizers.

Multiple risk factors

Global CHD risk is emphasized in reaching management decisions. Risk factors:

- are associated with disease
- are not necessarily causative
- should not be judged in isolation, or by number per se
- contribute to a global risk

A **mild** risk factor does not imply **mild** risk and Type 2 diabetes is not **mild** diabetes.

Multiplicative risk

Risk factors do not act in isolation; they may be additive or multiplicative. Thus, 40-year-old male smokers in the Framingham Study, who had marked hypertension, hypercholesterolaemia and glucose intolerance, had a 45% chance of CHD by the age of 48 years (Figure 14). Risk factors cluster together, and the threshold for intervention is lower in diabetes.

Treatment interactions

Increased risk is present in diabetes but commonly used hypertension treatments may aggravate existing metabolic disturbances (Table 12).

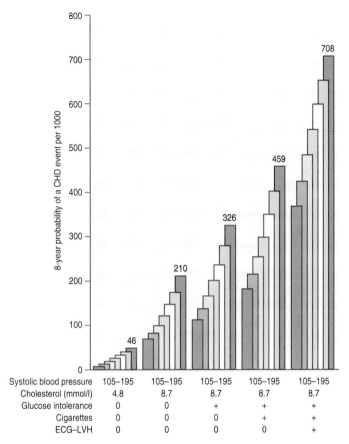

Figure 14

Cumulative risks of coronary heart disease over 8 years from increasing systolic blood pressure, in presence or absence of other risk factors, in 40-year-old Framingham Study men (at the 18-year follow up). Data from Houston.[88] (ECG–LVH, left ventricular hypertrophy on electrocardiogram.)

β-Blockers

β-Blockers may worsen the lipid profile (Table 12), raising triglyceride levels and lowering HDL levels. Direct evidence that this alters outcome is partly lacking, but reduction in CHD

Table 12
Effects of antihypertensive drugs on serum lipids and lipoproteins

	Serum cholesterol	Serum triglycerides	LDL cholesterol	HDL cholesterol	Cholesterol: HDL cholesterol
β-Blockers	→	↑↑↑	→	↓↓	↑
Thiazides	↑	↑↑	↑	↓	↑
ACE inhibitors	→	→	→	→	→
Calcium channel antagonists	→	→	→	→	→
α-Blockers	↓	↓↓	↓	↑	↓

in hypertension trials using these agents was somewhat disappointing when compared with reduction in stroke. β-Blockers and ACE inhibitors performed well and equally in Type 2 diabetes patients with hypertension in the UKPDS.[58–60] β-Blockade should be used when it is necessary for angina, and in the first year at least following myocardial infarction. β-Blockade can be beneficial in congestive cardiac failure, which is a particular problem in diabetes. Drugs with intrinsic sympathomimetic activity may be more lipid neutral. β-Blockers can reduce hypoglycaemia awareness in Type 1 diabetes, and delay the rate of recovery from hypoglycaemia but it is not a common clinical problem. β-Blockade may worsen symptoms of PVD.

Diuretics

Thiazides can elevate triglyceride, cholesterol and LDL levels, and lower HDL levels, exacerbating atheroma risk, but this is less marked in low dose. Loop diuretics are less dyslipidaemic in their effects. Thiazides may exacerbate Type 2 diabetes. In newly diagnosed Type 2 diabetes, whether or not there is also dyslipidaemia, thiazide withdrawal should be attempted. Thiazides are associated with impaired potency, which is a particular issue in diabetes.

Calcium channel blockers

Calcium channel antagonists are lipid neutral, and appropriate for dyslipidaemic individuals. Significant effects on insulin secretion and glucose tolerance are not usually seen. There is some evidence that they may be less effective agents than ACE inhibitors in Type 2 diabetic patients with proteinuria.[89,90]

Angiotensin-converting enzyme (ACE) inhibitors

ACE inhibitors are lipid neutral, reduce insulin resistance and are appropriate in dyslipidaemic hypertension, although a diuretic may be needed for the full hypotensive effect. Initial use of ACE inhibitors must be monitored for the acute deterioration in renal function that may occasionally occur when there is subclinical renal artery stenosis; this is particularly important in diabetes where renal artery stenosis is more common, especially if PVD is also present. Renal function should be checked before, one week after and one month after the introduction of an ACE inhibitor. ACE inhibitors are protective against progression of proteinuria and diabetic renal disease in Type 1 diabetes and probably in Type 2 diabetes due to (and perhaps also independent of) their hypotensive effect. As mentioned above, ACE inhibitors (captopril) and β-blockers (atenolol) performed similarly over 10 years in the UKPDS.[58–60]

α-Blockers

α-Adrenergic blockers are useful either alone or in combination for hypertension in the diabetic and/or dyslipidaemic patient. They improve the lipid profile (see Table 12), insulin resistance may decrease, and LPL and LDL-receptor activities increase.

Severe hypertriglyceridaemia

Introduction

Moderate hypertriglyceridaemia is common, but more severe hypertriglyceridaemia (≥ 5 mmol/l) is less so. Values over 10 mmol/l occur in about 0.2% of the population. Combinations of acquired and genetic factors cause over-synthesis of lipoproteins and/or impaired lipoprotein clearance (Table 13). Secondary risk factors for hypertriglyceridaemia include diabetes, obesity, alcohol, myxoedema, oestrogen, renal disease and drugs (Table 14).[91]

Table 13
Characteristic serum triglyceride concentrations in different conditions

	Serum triglyceride (mmol/l)
Lipoprotein lipase deficiency	≥ 20
Familial hypertriglyceridaemia and diabetes mellitus	≥ 10
Familial hypertriglyceridaemia	3–7
Uncontrolled diabetes mellitus (without other condition)	2–6

Table 14

Hypertriglyceridaemia-associated features

- LDL may be raised (eg in familial combined hyperlipidaemia)
- HDL cholesterol level often reduced
- With dysbetalipoproteinaemia (Type III, remnant lipaemia)
- With hyperchylomicronaemia
- With secondary disorders such as:
  diabetes mellitus
  ethanol
  obesity (especially central obesity)

Presentation

Abdominal pain occurs in some individuals and can progress to acute pancreatitis. Triglyceride levels are usually over 20 mmol/l, but may be missed if triglyceride levels settle quite quickly with intravenous fluids and negative calorie balance, prior to any lipid measurements. Recurrent pancreatitis can be minimized in these patients by control of triglycerides. Diabetes with severe hypertriglyceridaemia may lead to pancreatitis less frequently than when alcohol excess contributes to the pancreatitis. Hepatomegaly and splenomegaly are quite frequent, even without alcohol, with the triglyceride-rich lipoproteins being removed by reticuloendothelial cells.

Milky plasma (VLDL) is found and, on standing, plasma can develop a creamy surface (chylomicrons). Eruptive xanthomata (Figure 15) over the back, buttocks and limbs, consisting of raised papules that are usually non-itchy with yellowish centres, develop quite quickly, and resolve over weeks as triglyceride levels fall.

Severe hypertriglyceridaemia may be visible in the retinal vessels (lipaemia retinalis), but it is usually found after hypertriglyceridaemia has been diagnosed!

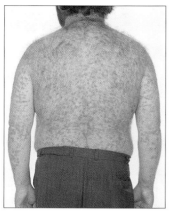

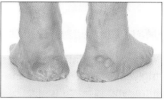

Figure 15
Eruptive xanthomata in severe hypertriglyceridaemia. Eruptive xanthomata are shown in a 38-year-old man of weight 110 kg and height 5 foot 10 inches, caused by severe hypertriglyceridaemia (cholesterol 17.0 mmol/l, triglycerides 88 mmol/l) in the presence of Type 2 diabetes (fasting blood glucose 8.8 mmol/l, fructosamine 315 µmol/l (normal 190–285)). Dietary efforts led to 4 kg weight loss, but triglyceride levels remained elevated at 70 mmol/l (cholesterol 16.0 mmol/l), and were unresponsive to fibrates. Insulin was given (1.2 units/kg), without hypoglycaemia, and adequate lipid control was obtained (cholesterol 5–6.5 mmol/l and triglycerides 5.6–12.0 mmol/l), sufficient to minimize risk of acute pancreatitis, and sufficient for these xanthomata to resolve.

Mild confusion, memory impairment, paraesthesiae and transient neurological problems have been reported, perhaps reflecting changes in viscosity and blood flow.

Laboratory investigations

In acute pancreatitis severe hypertriglyceridaemia should be considered. Severe hypertriglyceridaemia can interfere with serum amylase or any laboratory blood measurement based on colorimetry, giving spuriously low results. Amylase measurement may be possible after serum dilution, but urine amylase can always be measured.

Severe hypertriglyceridaemia takes up a significant plasma volume, with resultant difficulty with any other volume-based measurements. Apparent plasma sodium levels of less then 130 mmol/l can be found, when the real sodium level is over 150 mmol/l, and inappropriate treatment of pseudohyponatraemia with hypertonic saline may worsen undiagnosed true hypernatraemia.

Management of severe hypertriglyceridaemia

When abdominal pain or pancreatitis is present, lowering of the triglyceride levels is urgent. Triglycerides rapidly settle with intravenous fluids and calorie restriction. Longer-term weight loss is important, fat intake should be limited, alcohol intake should cease and diabetes requires treatment. Secondary conditions should be excluded or treated. Sometimes, diabetic individuals have plasma triglyceride levels that are elevated out of proportion to glucose; in this instance insulin treatment can be appropriate, although doses may be high as a result of insulin resistance. Rarely, plasma exchange using a cell separator has been used to lower triglyceride levels rapidly, alleviating abdominal symptoms.[92]

References

1. Pirart J. Diabetes mellitus and its degenerative complications. A prospective study of 4400 patients observed between 1947 and 1973. *Diabetes Care* 1978; **1**: 168–88, 252–63.

2. Diabetes Control and Complications Trial Research Group. The effect of intensive treatment of diabetes on the development and progression of long-term complications in insulin dependent diabetes mellitus. *N Engl J Med* 1993; **329**: 977–86.

3. UK Prospective Diabetes Study (UKPDS) Group. Intensive blood-glucose control with sulphonylureas or insulin compared with conventional treatment and risk of complications in patients with type 2 diabetes (UKPDS 33). *Lancet* 1998; **352**: 837–53.

4. UKPDS Group. Effect of intensive blood-glucose control with metformin on complications in overweight patients with type 2 diabetes (UKPDS 34). *Lancet* 1998; **352**: 854–65.

5. Fuller JH, Shipley MJ, Rose G et al. Coronary heart disease risk and impaired glucose tolerance: the Whitehall study. *Lancet* 1980; **i**: 1373–6.

6. Singer DE, Nathan DM, Anderson KM et al. Association of HbA_{1c} with prevalent cardiovascular disease in the original cohort of the Framingham Heart Disease Study. *Diabetes* 1992; **141**: 202–8.

7. Wilson PWF, Cupples LA, Kannel WB. Is hyperglycaemia associated with cardiovascular disease? The Framingham Heart Study. *Am Heart J* 1992; **121**: 586–90.

8. Marble A. Insulin – clinical aspects: the first fifty years. *Diabetes* 1972; **21**: 632–6

9. Sasaki A, Kamado K, Horiuchi N. A changing pattern of causes of death in Japanese diabetics: observations over fifteen years. *J Chronic Dis* 1978; **31**: 433–44.

10. Kawate R, Yamakido M, Nishimoto Y et al. Diabetes mellitus and its vascular complications in Japanese migrants on the island of Hawaii. *Diabetes Care* 1979; **2**: 161–70.

11. Fuller JH, Elford J, Goldblatt P, Adelstein AM. Diabetes mortality: new light on an underestimated public health problem. *Diabetologia* 1983; **24**: 336–41.

12. Reckless JPD. The epidemiology of heart disease in diabetes mellitus. In: Taylor KG, ed, *Diabetes and the heart*. Tunbridge Wells: Castle House Publications, 1987: 1–18.

13. Marks HH, Krall LP. Onset, course, prognosis and mortality in diabetes mellitus. In: Marble A, White P, Bradley RF, Krall LP, eds. *Joslin's Diabetes Mellitus*, 11th edn. Philadephia: Lea and Febiger, 1971: 209–54.

14. Zdanov VS, Vihert AM. Atherosclerosis and diabetes mellitus. *Bull* WHO 1976; **53**: 547–53.

15. Kannel WB, McGee DL. Diabetes and cardiovascular disease. The Framingham Study. *JAMA* 1979; **241**: 2035–8.

16. Kannel WB, McGee DL. Diabetes and glucose tolerance as risk factors for cardiovascular disease. The Framingham Study. *Diabetes Care* 1979; **2**: 120–6.

17. Most RS, Sinnock P. The epidemiology of lower extremity amputations in diabetic individuals. *Diabetes Care* 1983; **6**: 87–91.

18. Pyorala K, Laakso M, Uusitupa M. Diabetes and atherosclerosis: an epidemiological view. *Diabetes Metab Rev* 1987; **3**: 464–524.

19. Stamler J, Vaccaro O, Neaton JD, Wentworth D. Diabetes, other risk factors, and 12 year cardiovascular mortality for men screened in the Multiple Risk Factor Intervention Trial. *Diabetes Care* 1993; **16**: 434–44.

20. Groeneveld Y, Petri H, Hermans J, Springer MP. Relationship between blood glucose level and mortality in Type 2 diabetes mellitus: a systematic review. *Diabetic Med* 1999; **16**: 2–13.

21. Frick MH, Elo O, Haapa K et al. Helsinki Heart Study; primary prevention trial with gemfibrozil in middle-aged men with dyslipidaemia. *N Engl J Med* 1987; **317**: 1237–45.

22. Koskinen P, Manttari M, Manninen V et al. Coronary heart disease incidence in NIDDM patients in the Helsinki Heart Study. *Diabetes Care* 1992; **15**: 820–5.

23. Rubins HB, Robins SJ, Collins D et al. Gemfibrozil for the secondary prevention of coronary heart disease in men with low levels of high-density lipoprotein cholesterol. Veterans' Affairs High Density Lipoprotein Cholesterol Intervention Trial (VA-HIT) Study Group. *N Engl J Med* 1999; **341**: 410–18.

24. West KM, Ahuja MM, Bennett PH et al. The role of circulating glucose and triglyceride concentrations and their interactions with other 'risk factors' as determinants of arterial disease in nine diabetic population samples from the WHO Multinational Study. *Diabetes Care* 1983; **6**: 361–9.

25. Reckless JPD, Betteridge DJ, Wu P et al. High density and low density lipoproteins and prevalence of vascular disease in diabetes mellitus. *BMJ* 1978; **1**: 883–6.

26. Borch-Johnsen K, Kreiner S. Proteinuria: value as a predictor of cardiovascular mortality in insulin dependent diabetes mellitus. *BMJ* 1987; **294**: 1651–4.

27. Mogensen CE. Microalbuminuria predicts clinical proteinuria and early mortality in maturity-onset diabetes. *N Engl J Med* 1984; **310**: 356–60.

28. Lam KSL, Cheng IKP, Janus ED, Pang RWC. Cholesterol-lowering therapy may retard the progression of diabetic nephropathy. *Diabetologia* 1995; **38**: 604–9.

29. Steinberg D, Parthasarawathy S, Carew TE et al. Beyond cholesterol: modifications of low-density lipoprotein that increase its atherogenicity. *N Engl J Med* 1989; **320**: 915–24.

30. Graham IM, Daly LE, Refsum HL et al. Plasma homocysteine as a risk factor for vascular disease: the European Concerted Action Project. *JAMA* 1997; **277**: 1775–81.

31. Goldstein JL, Brown MS. Familial hypercholesterolaemia. In: Stanbury JB, Wyngaarden JB, Fredrickson DS, Goldstein JL, Brown MS, eds, *The Metabolic Basis of Inherited Disease*, 5th edn. New York: McGraw-Hill, 1983; 672–712.

32. World Health Organization. *Diabetes Mellitus: Report of WHO study group*. WHO, Geneva Switzerland 1985; WHO technical report series No. 727.

33. Report of the Expert Committee on the Diagnosis and Classification of Diabetes Mellitus. *Diabetes Care* 1998; **21**: S5–S19.

34. Kissebah AH, Alfarsi S, Evans DJ, Adams PW. Integrated regulation of very low density lipoprotein triglyceride and apolipoprotein-B kinetics in non-insulin dependent diabetes mellitus. *Diabetes* 1982; **31**: 217–25.

35. Lloyd J, Reckless JPD. Changes in lipid and lipoprotein concentrations in a community population with diabetes compared to matched non-diabetic controls. Unpublished data.

36. Howard BV. Lipoprotein metabolism in diabetes mellitus. *J Lipid Res* 1987; **28**: 613–28.

37. Pietri AO, Dunn FL, Grundy SM, Raskin P. The effect of continuous subcutaneous insulin infusion on very low density lipoprotein triglyceride metabolism in type 1 diabetes mellitus. *Diabetes* 1983; **32**: 75–81.

38. Winocour PH, Durrington PN, Ishola M, Anderson DC. Lipoprotein abnormalities in insulin dependent diabetes mellitus. *Lancet* 1986; **i**: 1176–8.

39. Fielding CJ, Reaven GM, Liu G, Fielding PE. Increased free cholesterol in plasma low and very low density lipoproteins in non-insulin dependent diabetes mellitus: its role in the inhibition of cholesterol ester transfer. *Proc Natl Acad Sci USA* 1984; **81**: 2512–16.

40. James RW, Pometta D. Differences in lipoprotein subfraction composition and distribution between type 1 diabetic men and control subjects. *Diabetes* 1990; **39**: 1158–64.

41. James RW, Pometta D. The distribution profiles of very low density and low density lipoproteins in poorly controlled male, type 2 (non-insulin dependent) diabetic patients. *Diabetologia* 1991; **34**: 246–52.

42. Austin MA, Breslow JL, Hennekens CH et al. Low density lipoprotein subclass patterns and risk of myocardial infarction. *JAMA* 1988; **260**: 1917–21.

43. Simpson HC, Williamson CM, Olivecrona T et al. Postprandial lipaemia, fenofibrate and coronary artery disease. *Atherosclerosis* 1990; **85**: 193–202.

44. Witztum JL. Role of oxidised low density lipoprotein in atherogenesis. *Br Heart J* 1993; **69 (suppl 1)**: S12–S18.

45. Hunt JV, Bottoms MA, Mitchinson MJ. Oxidative alterations in the experimental glycation model of diabetes mellitus are due to protein–glucose adduct oxidation. Some fundamental differences in proposed mechanisms of glucose oxidation and oxidant production. *Biochem J* 1993; **291**: 529–35.

46. Fuller JH, Shipley MJ, Rose G et al. Mortality from CHD and stroke in relation to degree of hyperglycaemia: the Whitehall Study. *BMJ* 1983; **287**: 867–70.

47. Lyons TJ. Glycation and oxidation: a role in the pathogenesis of atherosclerosis. *Am J Cardiol* 1993; **71**: 26B.

48. Steinberg D. Modified forms of low density lipoprotein and atherosclerosis. *J Intern Med* 1993; **223**: 227–32.

49. Rimm EB, Stampfer MJ, Ascherio A et al. Vitamin E consumption and the risk of coronary disease in men. *N Engl J Med* 1993; **328**: 1450–6.

50. Stampfer MJ, Hennekens CH, Manson J-AE et al. Vitamin E consumption and the risk of coronary disease in women. *N Engl J Med* 1993; **328**: 1444–9.

51. Salonen JT, Nyyssonen K, Korpela H et al. High stored iron levels are associated with excess risk of myocardial infarct in eastern Finnish men. *Circulation* 1992; **86**: 803–11.

52. Salonen JT, Yla-Herttuala, Yamamoto R et al. Autoantibody against oxidised LDL and progression of carotid atherosclerosis. *Lancet* 1992; **339**: 883–7.

53. Pyorala K, De Backer G, Graham I et al. Prevention of coronary heart disease in clinical practice. Recommendations of the task force of European Society of Cardiology, European Atherosclerosis Society and the European Society for Hypertension. *Eur Heart J* 1994; **15**: 1300–31.

54. Core Services Committee. Guidelines for management of mildly raised blood pressure in New Zealand. Core Services Committee 1994, pp1–20. PO Box 5013, Wellington, NZ. ISBN 0-477-01740-1.

55. Wood D, Durrington P, Poulter N et al.on behalf of the British Cardiac Society, British Hyperlipidaemia Association, British Hypertension Society and British Diabetic Association. Joint British recommendations on prevention of coronary heart disease in clinical practice. *Heart* 1998; **80 (suppl 2)**: S1–S29.

56. The ASPIRE Steering Group. A British Cardiac Society survey of the potential for the secondary prevention of coronary disease; ASPIRE (Action on Secondary Prevention through Intervention to Reduce Events). *Heart* 1996; **75**: 334–42.

57. Reckless JPD. The economics of cholesterol lowering. *Clin Endocrinol Metab* 1990; **4**: 947–72.

58. UKPDS Group. Tight blood pressure control and risk of macrovascular and microvascular complications in type 2 diabetes: UKPDS 38. *BMJ* 1998; **317**: 703–13.

59. UKPDS Group. Efficacy of atenolol and captopril in reducing risk of macrovascular and microvascular complications in type 2 diabetes: UKPDS 39. *BMJ* 1998; **317**: 713–20.

60. UKPDS Group. Cost effectiveness analysis of improved blood pressure control in hypertensive patients with type 2 diabetes: UKPDS 40. *BMJ* 1998; **317**: 720–6.

61. Betteridge DJ, Dodson PM, Durrington PN et al. Management of hyperlipidaemia: guidelines of the British Hyperlipidaemia Association. *Postgrad Med J* 1993; **69**: 359–69.

62. Caro J, Klittich W, McGuire A et al. for the West of Scotland Coronary Prevention Study Group. The West of Scotland coronary prevention study: economic benefit analysis of primary prevention with pravastatin. *BMJ* 1997; **315**: 1577–82.

63. Haq IU, Jackson PR, Yeo WW, Ramsay LE. Sheffield risk and treatment table for cholesterol lowering for primary prevention of coronary heart disease. *Lancet* 1995; **346**: 1467–71.

64. Ramsey LE. Target lipid-lowering drug therapy for primary prevention of coronary disease: an updated Sheffield table. *Lancet* 1996; **348**: 387–8.

65. Winyard G. SMAC (Standing Medical Advisory Committee) statement on the use of statins. Department of Health, London 1997: EL(97)41, August 1997.

66. Department of Health (UK). National Service Framework for Coronary Heart Disease: modern standards and service models. Department of Health, London 2000.

67. Reckless JPD. Cost-effectiveness of statins. *Curr Opin Lipidol* 2000 (in press).

68. Nutrition Subcommittee, British Diabetic Association. Dietary recommendations for diabetics for the 1980s –- A policy statement. *Human Nutrition: Applied Nutrition* 1982; **36A**: 378–94.

69. Kinmouth AL, Macgrath G, Reckless JPD et al (Nutrition Subcommittee, British Diabetic Association). Dietary recommendations for children and adolescents with diabetes. *Diabetic Med* 1989; **6**: 537–47.

70. Howard BV, Abbott WGH, Swinburn BA. Evaluation of metabolic effects of substitution of complex carbohydrates for saturated fat in individuals with obesity and non-insulin dependent diabetes mellitus. *Diabetes Care* 1991; **14**: 786–95.

71. Parfitt VJ, Desomeaux K, Bolton CH, Hartog M. Effects of high monounsaturated and polyunsaturated fat diets on plasma lipoproteins and lipid peroxidation in type 2 diabetes mellitus. *Diabetic Med* 1994; **11**: 85–91.

72. Stone DB, Connor WE. The prolonged effects of a low cholesterol, high carbohydrate diet on the serum lipids in diabetic patients. *Diabetes* 1963; **12**: 127–32.

73. Reckless JPD. Dietary fibre and blood lipids. In: *Dietary fibre in the management of the diabetic*. Oxford: Medical Education Services, for the British Diabetic Association, 1984: 15–20.

74. Sjöstrröm L, Rissanen A, Andersen T et al. for the European Multicentre Orlistat Study Group. Randomised placebo-controlled trial of orlistat for weight loss and prevention of weight regain in obese patients. *Lancet* 1998; **352**: 167–72.

75. Alberti KGMM, Gries F. Management of non-insulin dependent diabetes in Europe: a consensus statement. *Diabetic Med* 1988; **5**: 275–81.

76. British Hyperlipidaemia Association. Fact Sheet 4: Diabetes Mellitus: Vascular disease and management of dyslipidaemia. British Hyperlipidaemia Association 2000; BHA, c/o DMC, AG 7&8, Aston Science Park, Birmingham B7 4BJ, UK.

77. The Scandinavian Simvastatin Survival Study Group. Randomised trial of cholesterol lowering in 4444 patients with coronary heart disease: the Scandinavian Simvastatin Survival Study (4S). *Lancet* 1994; **344**: 1383–9.

78. Sacks FM, Pfeffer MA, Lemuel AM et al. The effect of pravastatin on coronary events after myocardial infarction in patients with average cholesterol levels. *N Engl J Med* 1996; **335**: 1001–9.

79. The long-term intervention with pravastatin in ischaemic disease (LIPID) study group. Prevention of cardiovascular events and death with pravastatin in patients with coronary heart disease and a broad range of initial cholesterol levels. *N Engl J Med* 1998; **339**: 1349–57.

80. Shepherd J, Cobbe MS, Ford I et al. Prevention of coronary heart disease with pravastatin in men with hypercholesterolaemia. *N Engl J Med* 1995; **333**: 1301–7.

81. Downs JR, Clearfield M, Weis S et al. for the AFCAPS/TexCAPS Research Group. Primary prevention of acute coronary events with lovastatin in men and women with average cholesterol levels. *JAMA* 1998; **279**: 1615–22.

82. Reckless JPD. Cost-effectiveness of hypolipidaemic drugs. *Postgrad Med J* 1993; **69 (suppl 1)**: 30–3.

83. Farnier M, Dejager S and the French Fluvastatin Study Group. Effect of combined fluvastatin–fenofibrate therapy compared with fenofibrate monotherapy in severe hypercholesterolemia. *Am J Cardiol* 2000; **85**: 53–7.

84. Pyorala K, Pedersen TR, Kjekhus J et al. and the Scandinavian Simvastatin Survival Study (4S) Group. Cholesterol lowering with simvastatin improves prognosis of diabetic patients with coronary heart disease: a subgroup analysis of the Scandinavian Simvastatin Survival Study (4S). *Diabetes Care* 1997; **20**: 614–20.

85. Pedersen TR, Olsson AG, Faergeman O et al. Lipoprotein changes and reduction in the incidence of major coronary heart disease events in the Scandinavian Simvastatin Survival Study (4S). *Circulation* 1998; **97**: 1453–60.

86. Lipid Research Clinics Program. The Lipid Research Clinics coronary primary prevention trial results: reduction in incidence of coronary heart disease. *JAMA* 1984; **251**: 351–64.

87. Mansell P, Lloyd C, Stirling C et al. Effect of dried garlic tablets on serum lipids in diabetic patients. *Diabetic Med* 1994; **11 (suppl 1)**: 515.

88. Houston MC. Treatment of hypertension in diabetes mellitus. *Am Heart J* 1989; **118**: 819–29.

89. Tatti P, Pahor M, Byington RP et al. Outcome results of the Fosinopril versus Amlodipine Cardiovascular Events randomised trial (FACET) in patients with hypertension and NIDDM. *Diabetes Care* 1998; **21**: 597–603.

90. Califf RM, Granger CB. Commentary. Hypertension and diabetes and the Fosinopril versus Amlodipine Cardiovascular Events Trial (FACET): more ammunition against surrogate endpoints. *Diabetes Care* 1998; **21**: 655–7.

91. Reckless JPD. Isolated hypertriglyceridaemia. *Prescribers' Journal* 1996; **36**: 85–92.

92. Betteridge DJ, Bakowski M, Taylor KG et al. Treatment of severe diabetic hypertriglyceridaemia by plasma exchange. *Lancet* 1978; **ii**: 1368.

Index